The Theory of Gravity Propulsion

By

Larry James

ISBN: 1-4033-9423-7 (e-book)
ISBN: 1-4033-9424-5 (Paperback)
ISBN: 1-4033-9425-3 (Dust Jacket)

This book is printed on acid free paper.

Cover art by Gina Georgousis. You can contact her at cloud999@aei.ca.

1stBooks - rev. 02/07/03

Dedication

This book was made possible because of the love and support of my wife and mother. My wife, Martha James, encouraged me to pursue my dreams while she sacrificed much to support me in this effort. My mother, Peggie James, generously helped us financially while I worked to develop this theory. Without their unselfish support, this report would not exist.

Table of Contents

1.0 Introduction

This book explains a new theory called The Theory of Gravity Propulsion or TGP for short. TGP provides a conceptual approach to the Grand Unification Theory. It explains how all matter, energy, fields, forces, time and space works. Yes, it explains how gravity works too.

In 1915, Albert Einstein developed his Theory of General Relativity. His theory states that gravity forces are exactly equivalent to acceleration forces. In other words, gravity magically changes time and space. I call it magic because his theory does not explain how gravity does it trick to change time and space. This book reveals the magic.

TGP discovers and defines a new force called gravity propulsion. Gravity propulsion provides important benefits to us. It is gravity propulsion that holds our universe together. Without gravity propulsion, there would be no gravity, no planets, and no stars, no mass and no atoms. Time depends on gravity propulsion forces. Without gravity propulsion, time would stop. However, gravity propulsion creates challenges to our existence. It is gravity propulsion that is expanding the universe and is causing it to accelerate its expansion. And, gravity propulsion is rapidly leaving us. It is gravity propulsion that is causing our universe to end.

We are living in the midst of a powerful explosion that is beyond comprehension. The force around us is more powerful than the explosions of our Sun. Wouldn't it be nice if we could control this force that is around us? This report shows how we can do just that. We can increase gravity, reduce gravity, reverse gravity, speed up time, slow down time and more. This is just the beginning of the wonderful things we can now do.

I am asking you to consider some very unorthodox concepts. Up until 1948, many experts thought airplanes could never go faster than the speed of sound. Today, there are experts who believe nothing can go faster than the speed of light. TGP predicts that quantum strings travel faster than light. We show how to prove this. TGP predicts that we can communicate faster than light. We show how to do this.

The following is an overview of the main new discoveries and predictions that TGP makes.

Gravity

We will explain how gravity works and how to control it. TGP reveals discoveries of two gravity fields called *mass fields* and *gravity propulsion fields*. We will first explain how gravity is a result of interactions among these fields. Later we will explain how gravity is also a result of interactions among quantum strings that create these fields.

We will explain the discovery of the delta gravity equation:

$$\text{Change of Gravity} = D_t - D_a - I$$

We can control these variables and thereby we can increase, reduce and reverse gravity. We will define simple experiments where we can do this.

Time

We will explain how time works and how to control it. TGP reveals the discovery that the speed of time is directly proportional to the gravity propulsion force. We will explain the discovery of the time equation:

$$\text{Time} = 1 + (D_t - D_a + I - G)/P_o$$

We can control the variables D_t, D_a and "I" and thereby we can control the speed of time. We will show how we can do this.

We will explain a major discovery, "Our speed of time is slowing down." This also means that our speed of light is speeding up. We will show how to prove this.

The Universe

TGP reveals major new discovers and predictions about our universe. The main discoveries are described below.

- There are supporting facts that our universe began with a Big Bang. TGP goes beyond that. TGP explains for the first time what caused the Big Bang.
- TGP reveals that there is a center to our universe. TGP gives supporting facts and shows the direction to the center of our universe.
- TGP discovered that the speed of time is faster towards the center of our universe. TGP gives amazing facts that support this.

- There are supporting facts that our universe is expanding and its expansion is accelerating. TGP reveals that it is the omnipresent gravity propulsion force that causes this expansion and its acceleration.
- TGP predicts there is an invisible universe around us that is full of invisible stars and invisible galaxies radiating invisible light. This invisible universe is far larger than the one we can see. We will explain supporting facts.
- TGP explains for the first time why there are huge voids in our universe.
- TGP explains how the super cluster of galaxies was formed.
- TGP explains why the Hubble constant is not a constant.
- TGP predicts that we are permanently losing matter every second.
- TGP predicts that our universe will end.

The Unified Field Theory

TGP reveals the discovery of the Unified Field Theory. It explains how all fields, forces and energy works. TGP explains how one type of quantum string creates all forces including:

- Gravity forces
- Electrostatic forces
- Magnetic forces
- Electromagnetic forces
- Nuclear strong forces (What holds the nucleus together)
- Nuclear weak forces (Why nuclei decay)
- Gravity propulsion forces
- Inertia forces
- Centrifugal forces
- Particle magnetic pole forces
- All attractive and repulsion forces
- The forces that holds all particles together
- Lamb shift forces

The Grand Unification Theory

TGP goes beyond the Unified Field Theory. TGP also explains how all matter works including anti-matter, subatomic particles and invisible matter.

TGP explains how one type of quantum string creates all matter, fields, forces and energies. This is the Grand Unification Theory. This reveals new discoveries and predictions including the major prediction of the invisible universe.

The Theory of Gravity Propulsion

TGP goes beyond the expected Grand Unification Theory of matter and energy because it also explains time and space. TGP explains **the beginning and end** of time, space and the universe.

A New Field of Science

The Theory of Gravity Propulsion establishes a new field of science. It challenges existing laws. It explains events and facts that are otherwise unexplainable. It reveals new discoveries and makes new verifiable predictions. I hope this will stimulate research in this new field.

The first law of physics is, "There are no absolute laws of physics." "Laws of physics" seem like absolute facts that can never change. This is untrue. A "law of physics" actually means, "This is our best guess about how something works based on what we know so far." As we learn more, these laws can change. This Theory of Gravity Propulsion establishes a new foundation of physics. I am not proud of this theory's renegade aspect. I am proud of this theory's ability to provide a deeper understanding of the universe.

I hope this report inspires you to know more about the forces of the universe. I hope to bring physics out of the lab and into your personal life. When I was a little boy, my Dad took me outside at night. We looked up at the clear dark sky and gazed at the bright stars. We talked about how far away they were, how big the universe is and wondered how we fit in. 'Twinkle, twinkle little star. How I wonder what you are.' This child-like wonder has long inspired me to know more and more. I hope you will go outside on a clear night, gaze up at the stars and wonder what they are. I hope the awe and wonder of our vast and powerful universe that God created for you will touch you too.

This theory has opened a new exciting door of science. I hope you will enter in. There are many topics for people to research to advance this

4

theory. I have identified various concrete experiments that could and should be conducted. This book identifies areas that need research by physicists, scientists, astronomers, cosmologists, mathematicians and many other disciplines. Because this is new uncharted territory, you may make important discoveries and advancements. To help advance this theory, there is a web site dedicated to coordinate research in these areas, announce new discoveries and provide updates to this report. This is www.universalpower.org.

Predictions

As you explore these new ideas, we need to remind ourselves that only experiments establish scientific truth. Theory and math are tools we use to help us look in the right direction. This Theory of Gravity Propulsion predicts several new effects that can and should be experimentally measured to verify, disprove or refine this theory. This book describes ways to perform each of these experiments. These new verifiable predictions are listed in the table below along with the section that describes them.

No.	Verifiable Prediction	Section
1.	Reduced weight of a spinning object.	4.3.5
2.	Increased gravity around edges of a spinning object.	4.3.1/5
3.	Faster time around edges of a spinning object.	4.3.3/5
4.	Reduced gravity above a spinning object.	4.3.1/5
5.	Slower time above a spinning object.	4.3.3/5
6.	Reduced gravity below a spinning object.	4.3.1/5
7.	Slower time below a spinning object.	4.3.3/5
8.	Our speed of light is speeding up. Actually, our speed of time is slowing down.	7.5
9.	The direction towards the center of the universe has the highest average density of galaxies. The opposite direction has the lowest average density of galaxies.	7.12.1
10.	The direction towards the center of the universe has a supercluster of galaxies that appears to be the closest to us.	7.12.2
11.	The direction towards the center of the universe has the fastest speed of time. The opposite direction has the slowest speed of time.	7.12.3

12.	The direction towards the center of the universe has the highest average background radiation. The opposite direction has the lowest average background radiation.	7.12.4
13.	Light bends towards a negative electrostatic field.	9.6.2
14.	Light bends in a magnetic field as a positive particle.	9.6.3
15.	Invisible stars.	10.3
16.	The Lamb Shift effect.	10.6
17.	Communicate faster than light.	11.0
18.	Quantum strings travel faster than light.	11.3

2.0 The Gravity Propulsion Story

I will begin by telling you a short story about gravity propulsion. This story should help you understand about the life of a gravity propulsion field. As you continue to read this book, you may find it helpful to relate TGP to this story.

This story begins with an atom. The atom generated a mass field around it. As it generated its mass field, its mass was reduced. Then mass fields from another atoms combined with part of this atom's mass field and created a gravity propulsion field. Eventually, part of the gravity propulsion field slammed into another atom's mass. The force pushed the atom backward some and increased its mass some. And, the cycle continues. The End

This story has a happy ending. Sometimes it doesn't end like that. Sometimes gravity propulsion fields head for the edge of the universe never to be seen again. Figure 1 illustrates this story.

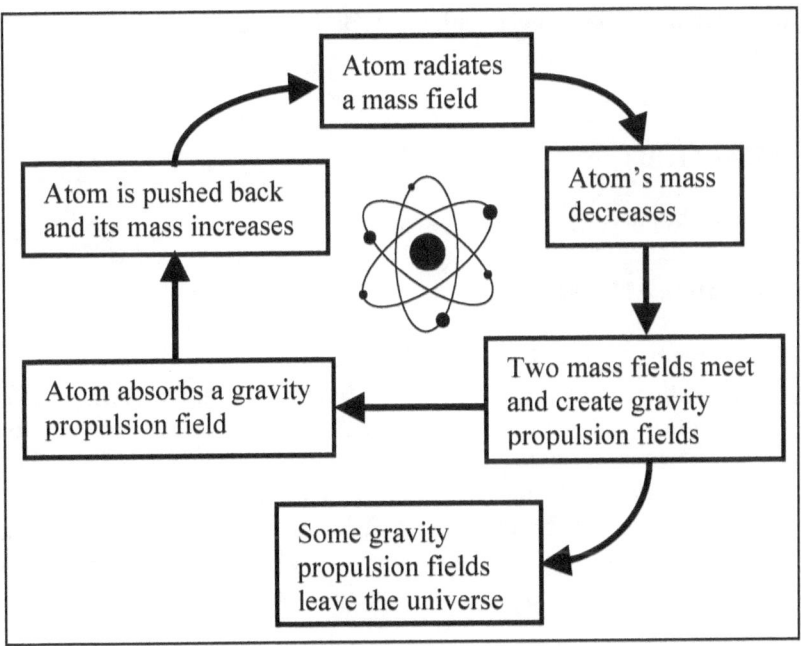

FIG. 1. Flow chart showing life cycle of gravity propulsion fields.

3.0 Gravity Fields

Gravity is a result between two types of gravity fields called *mass fields* and *gravity propulsion fields*. It is these two types of gravity fields that create gravity, gravity propulsion, inertia, centrifugal forces and also controls the speed of time.

Mass fields are gravity fields that radiate outward in an orderly manner from all atoms and masses (a collection of atoms). *Gravity propulsion fields* are gravity fields that are not as orderly. They are more random because they are resultant fields from interference with mass fields and/or other gravity propulsion fields. Mass fields become gravity propulsion fields after they interfere with (meet) other mass fields or other gravity propulsion fields.

Mass fields push outward on objects less than gravity propulsion fields push inward on them. The resultant force is gravity. Understanding these two fields will help you better understand the universe. It has also helped me to better appreciate the wonders of God. I hope it will help you too. These fields are described next.

3.1 Mass Fields

This section describes the properties of mass fields.

3.1.1 Creation of Mass Fields

This describes how mass fields are created.
1. All particles generate a mass field around them. All mass has a mass field around it. Mass fields are energy that radiate out from all particles.
2. Every atom's mass is reduced by the energy of the mass field it radiates. Section 3.2.3 explains that particle mass is increased by the energy of the gravity propulsion field it absorbs. This constant radiation and absorption provides a balance.

3.1.2 Propagation of Mass Fields

This describes the properties of mass fields as they travel.
1. Mass fields radiate outward from a mass in all directions.

2. Mass field's strength diminishes as distance increases from the center of the mass according to the formula $1/Distance^2$.
3. Mass fields can also be described as wave properties. All particles radiate a mass field with wave properties. Wave properties are related to the polarity and size of the mass.
4. Mass fields travel slightly faster than the speed of light. This unorthodox statement is defended in section 9.12 "Quantum Strings Travel Faster Than Light".

3.1.3 Effects of Mass Fields

This describes how mass fields affect other fields and masses and how other fields affect mass fields.

1. Mass fields are not affected by electromagnetic fields. Neither electricity, electrostatic nor magnetic fields have any affect on these fields.
2. Interaction of mass fields with other mass fields or other gravity propulsion fields creates gravity propulsion fields. When a mass field meets another mass field, it creates a gravity propulsion field. When a mass field meets a gravity propulsion field, it creates a stronger gravity propulsion field.
3. Mass fields push atoms in a mass outward less than incoming gravity propulsion fields push inward on the same mass. The resultant force is what we call gravity.

3.2 Gravity Propulsion Fields

This section describes the properties of gravity propulsion fields.

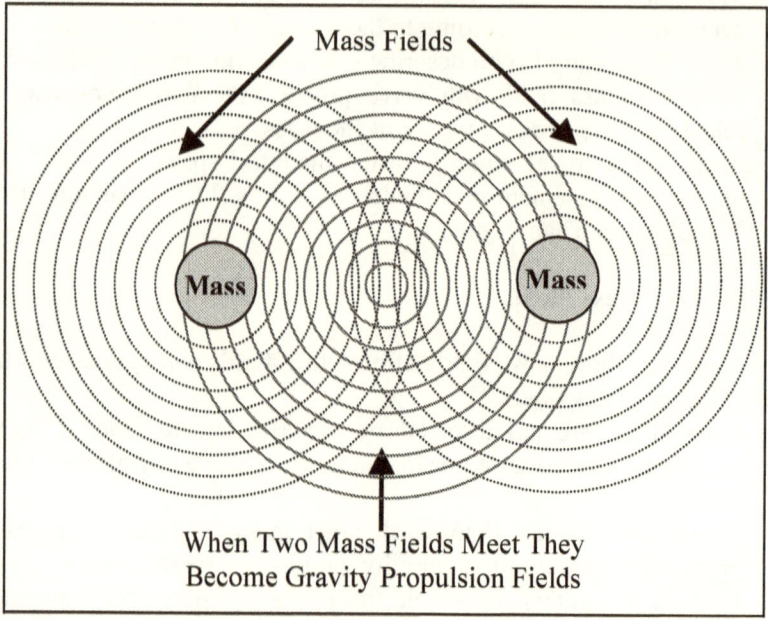

FIG. 2. Illustration of two mass fields creating a gravity propulsion field.

3.2.1 Creation of Gravity Propulsion Fields

This describes how gravity propulsion fields are created.

1. When a mass field meets another mass field, they create gravity propulsion fields.
2. When a mass field meets a gravity propulsion field, they create stronger gravity propulsion fields.

3.2.2 Propagation of Gravity Propulsion Fields

This describes the properties of gravity propulsion fields as they travel.

1. Gravity propulsion field's strength diminishes as distance increases and its strength increases as it interacts with other mass fields.
2. Gravity propulsion fields travel slightly faster than the speed of light. This unorthodox statement is defended in section 9.12 "Quantum Strings Travel Faster Than Light".

3.2.3 Effects of Gravity Propulsion Fields

This describes how gravity propulsions fields affect other things and are affected by other things.

1. Gravity propulsion fields affect all particles.
2. When a particle absorbs gravity propulsion fields it pushes the particle in the direction of the field.
3. The particle mass is increased by the energy of the gravity propulsion field it absorbs.
4. Gravity propulsion fields are deflected away from increasing gravity propulsion field intensity.
5. The speed of time is directly proportional to the intensity of gravity propulsion fields.

3.3 Deductive Logic

TGP is based on deductive logic. Here are examples of the first four deductive steps.

1. Gravity is a force that acts on all types of matter. Gravity depends on the total mass of the object and does not depend on the type of material that it is composed of. It doesn't matter if the material is magnetic or non-magnetic, conductive or not conductive, solid, liquid or gas, etc. Any theory about gravity must also explain all other similar forces that depend on the mass of the object and do not depend on the type of matter. There are only three forces that meet these requirements. These are; gravity, inertia and centrifugal forces. Therefore, any theory that explains gravity forces **must** also explain how inertia forces (called Newton's Laws of Motion) and centrifugal forces work.

2. Newton's First Law of Motion states. "An object at rest tends to remain at rest unless acted upon by an outside force". In other words, objects resist being accelerated. When an object is accelerated, a force pushes back on it. Therefore, this resistive force must be the result of the object being moved through a field. We will call this field X for now.

3. Newton's First Law of Motion also states, "An object in motion tends to continue in motion in a straight line unless acted upon by an outside force." When the object is in motion it must be moving

11

along with field X. Otherwise, as deduced in #2 above, field X would push back on the object and resist the object from being moved through it. Therefore, field X cannot come from any outside source or outside reference point. Field X must originate from the mass itself. That is why field X always moves along with the mass. We will call this a *mass field*. Therefore, we must conclude that all mass generates a *mass field* around it.

4. As stated in #2, "When an object is accelerated, a force pushes back on it." When an object is accelerated, its current mass field must move through some of its previous mass field. Therefore, the interaction (also called interference) of these two fields must create a new field and a new force that pushes back on it. We will call this new field the *gravity propulsion field*. When the gravity propulsion field meets an object, it pushes on the object. We will call this the *gravity propulsion force*.

With this these first four steps in deductive logic, we have identified the two gravity fields called *mass fields* and *gravity propulsion fields*. We have also concluded that gravity propulsions field are created by interference between two (or more) mass fields. We have also concluded that the gravity propulsion force pushes against matter. These are a few examples how deductive logic works.

This book just presents the main conclusions of TGP. It does not explain the many deductive steps in logic taken to reach these conclusions. That would probably fill several volumes and would probably be boring to read.

3.4 Questions

The questions in this book are to confirm that you understand important concepts presented so far. The following questions are about the Theory of Gravity Propulsion as stated in this section. The answers are in Section 16.

3.1 What happens to an atom when it radiates a mass field?
3.2 Do all atoms radiate a mass field all the time?
3.3 When two mass fields meet, what do they create?
3.4 When a mass field meets a gravity propulsion field, what do they create?
3.5 What two things happen to an atom when it encounters a gravity propulsion field?

4.0 Examples

Now that we have defined the properties of mass fields and gravity propulsion fields, we can explain their role in a few fascinating phenomena. We will explain:

1. Inertia
2. Centrifugal force
3. The Spinning Disk
4. Spinning projectiles (gun shells)
5. Searl's flying saucer.
6. Podkletnov's spinning disk.

After this, we will explain gravity using the same field properties.

4.1 Inertia

Let's first review what inertia is. In order to understand gravity, you have to first understand inertia. Sir Isaac Newton's first law of motion describes inertia this way; **"An object at rest tends to remain at rest, and an object in motion tends to continue in motion in a straight line unless acted upon by an outside force."**[1] Inertia is the property of matter that causes it to resist any change of its motion in either direction or speed. These inertia forces work the same way on earth as they do in the vacuum of outer space. Neither gravity forces, magnetic forces, nor electric forces affect inertia forces.

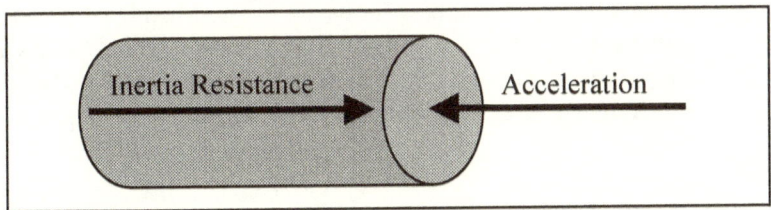

FIG. 3. Inertia resistance opposing acceleration.

Inertia produces a strong force that you can feel. Pick up a pen in your hand and shake it back and forth as fast as you can. Now pick up a heavier

---

[1] Microsoft® Encarta® Encyclopedia 99. © 1993-1998 Microsoft Corporation. All rights reserved.

object that weighs a few pounds that you can firmly hold in your hand. Shake it back and forth as fast as you can. You can feel it resisting your attempts to move it back and forth as fast as you did the lighter pen. With faster acceleration, the inertia resistance can be many times the weight of the object itself. What is this force? What causes this? What force causes an object to resist being moved? The answers reveal amazing secrets of the universe.

4.1.1 How Inertia Works

This section describes in more detail how mass fields and gravity propulsion fields create inertia resistance force. Gravity works in a similar manner. Compare this section to section 5.0.

Step 1: Mass fields. Remember, "All particles generate a mass field around them. Mass fields are energy that radiate out from all particles." (Section 3.1.1) The object radiates a mass field around it.

Step 2: Creation of gravity propulsion fields. Also recall that, "When a mass field meets another mass field, it creates a gravity propulsion field." (Section 3.1.3) When the object is accelerated, it moves through its own mass field. Its current mass field moves through some of its older mass field. This creates it's own gravity propulsion fields.

Imagine a section of a mass field as a loop around the object that is growing bigger. (See the following illustration, Figure 4.) As this expands, new loops are created inside of it and they expand too. When an object is accelerated, a new mass field loop moves into and meets an older mass field loop. The interference of these two mass fields creates gravity propulsion fields.

Step 3: Inertia is the resultant force. The resultant gravity propulsion fields are slightly tilted instead of being parallel to the line of acceleration. More gravity propulsion fields are pushing the mass inward and against the direction of acceleration. And more gravity propulsion fields are leaving outward in the direction of acceleration and do not push on the mass. The object's atoms will absorb more of the inward tilted gravity propulsion fields than it absorbs of the outward tilted gravity propulsion fields in the opposite direction. This results in a push backwards that we call inertia resistance.

One way to visualize the resultant gravity propulsion fields is to imagine a cone shape. The tip of the cone is in the middle axis of the object pointing against the direction of acceleration. The top of the cone is outside and around the object in the direction of the acceleration. This represents the resultant push, which is against the direction of acceleration.

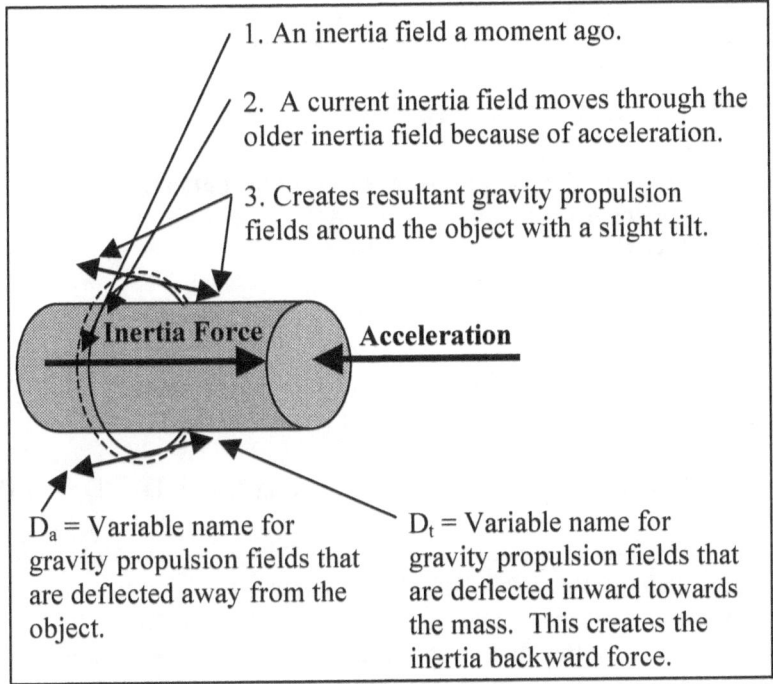

1. An inertia field a moment ago.

2. A current inertia field moves through the older inertia field because of acceleration.

3. Creates resultant gravity propulsion fields around the object with a slight tilt.

Inertia Force   **Acceleration**

D_a = Variable name for gravity propulsion fields that are deflected away from the object.

D_t = Variable name for gravity propulsion fields that are deflected inward towards the mass. This creates the inertia backward force.

FIG. 4. Shows how acceleration creates resultant opposing gravity propulsion forces and fields D_a and D_t.

I hope you appreciate inertia better now. Understanding how inertia works is the key to understanding gravity, gravity propulsion and many wonders of the universe.

4.1.2 TGP's Second Law of Motion

The classical equation, for Newton's Second Law of Motion, is F = MA. Newton's Law states that a force (F) is equal to a mass (M) times its acceleration (A). TGP modifies this equation to:

F = MD

Where:
- F = Force.
- M = Mass of the object.
- D = Resultant deflected gravity propulsion fields.

TGP's Second Law of Motion is,

Inertia force is the result of gravity field deflections that were created by accelerating a mass. Inertia force equals mass times the resultant deflected gravity propulsion fields.

Inertia force is the result of interactions among gravity fields that were created by accelerating a mass. Acceleration of a mass deflects gravity propulsion fields inward/toward and outward/away from the mass. It is these resultant fields acting on the mass that create inertia forces.

For objects that accelerate in a straight line: $A = D = D_t + D_a$ and $D_t = D_a$ where:
- A = Acceleration of the object.
- D_t = Resultant gravity propulsion fields that are deflected towards the object. D_t means "Deflected Towards". These deflected fields push back on the object against its direction of acceleration.
- D_a = Gravity propulsion fields that are deflected away from the object. D_a means "Deflected Away". These deflected fields no longer push forward on the object in its direction of acceleration.

This equation works for relatively slow accelerations. Extremely fast accelerations have additional contributing factors such as increasing mass.

4.2 Centrifugal Force

Centrifugal force is the name of the force with which a body, moving around a center, tends to fly off from that center. According to TGP, centrifugal forces and inertia forces are the same forces.

Figure 5 below is a top view of a spinning disk. Think of it as a collection of atoms. As the atoms rotate around, they are being accelerated toward the center by centripetal force. "Centripetal" means "towards the center". If the atoms were not being accelerated toward the center, each atom would travel in a straight line tangential to their rotation.

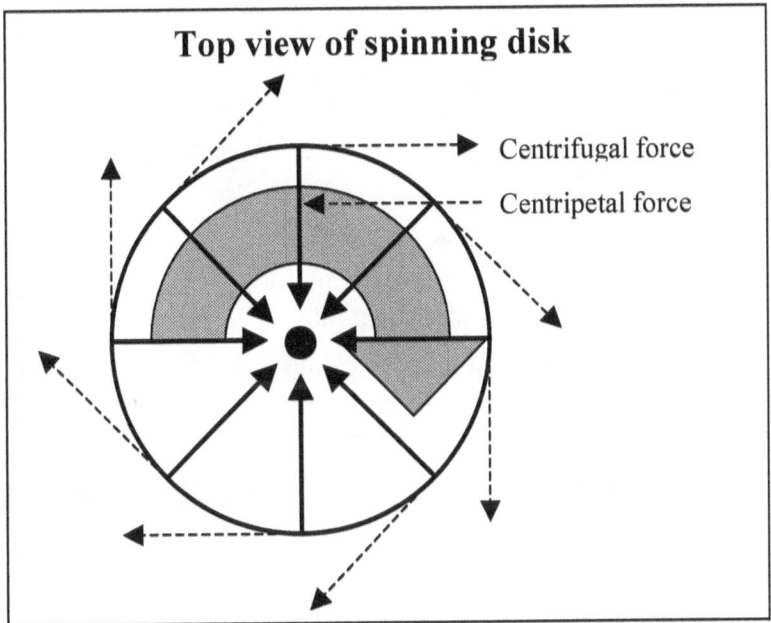

Top view of spinning disk

Centrifugal force
Centripetal force

FIG. 5. Centrifugal and centripetal forces of a spinning disk.

4.2.1 How Centrifugal Force Works

In any rotating object, each atom is being accelerated toward a center. As they are accelerated, they move through the object's older (an instant ago) mass fields. The interference of these two fields creates resultant gravity propulsion fields with a slight tilt. One resultant gravity propulsion field moves toward the center, the same direction the atom is accelerating, but with a slight tilt away from the object. The other resultant gravity propulsion field moves away from the center, in the opposite direction the atom is accelerating, but with a slight tilt inward toward the object. The object's atoms will absorb more of the inward tilted gravity propulsion fields than it absorbs of the outward tilted gravity propulsion fields. This results in an outward push that we call centrifugal force. See the following illustration.

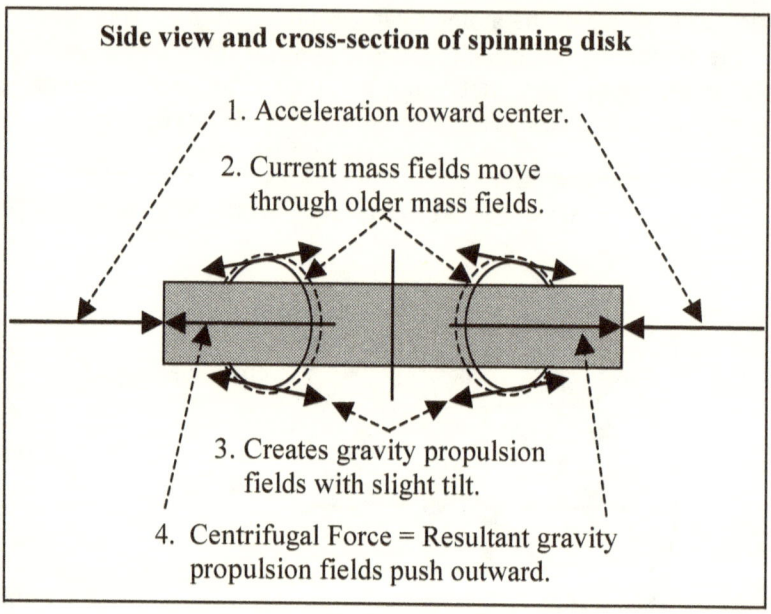

FIG. 6. Spinning disk creates gravity propulsion fields and centrifugal forces.

4.3 The Spinning Disk (Predictions 1 to 7)

An examination of the spinning disk provides the first exciting examples of how TGP can predict alterations in gravity and time. TGP predicts if you spin a cylinder fast enough, it will overcome the force of gravity and lift up. In other words, it will "fly". There are two effects that contribute to make it fly: Deflection Effect and Lift Effect. The Deflection Effect reduces the downward gravity propulsion force on an object. The Lift Effect increases the upward gravity propulsion force on an object. The Deflection Effect means gravity propulsion fields are deflected around the spinning object, which reduces its weight. The Lift Effect is an increased gravity propulsion force upward that contributes to lifting the spinning object. The Deflection

Effect reduces the cylinder's weight about twice as much as the Lift Effect lifts the cylinder. When the deflection and lift forces are greater than its normal weight, the spinning cylinder will travel up and away from the Earth.

As explained previously, as a disk spins, it generates gravity propulsion fields. This is because atoms are being accelerated towards the center.

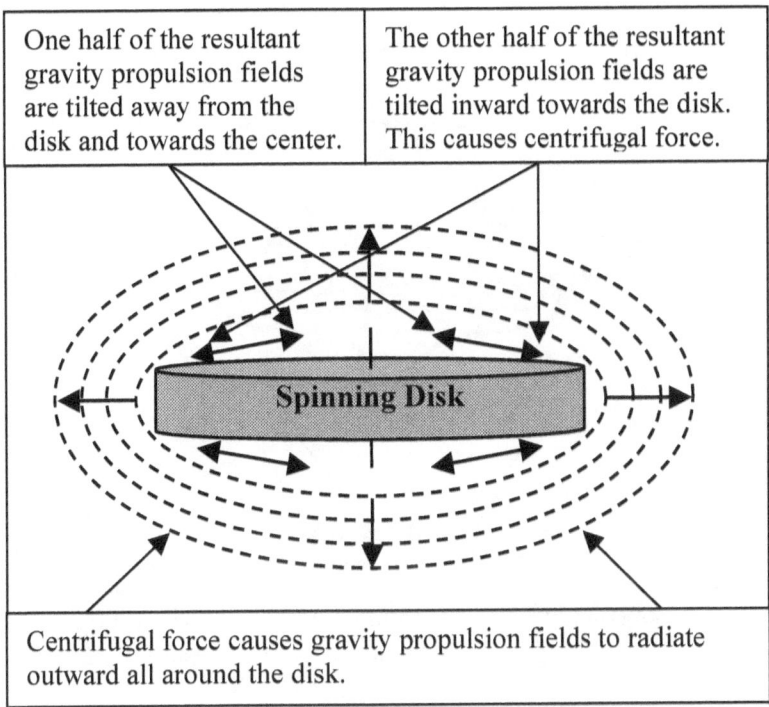

One half of the resultant gravity propulsion fields are tilted away from the disk and towards the center.	The other half of the resultant gravity propulsion fields are tilted inward towards the disk. This causes centrifugal force.

Centrifugal force causes gravity propulsion fields to radiate outward all around the disk.

FIG. 7. Shows spinning disk creates surrounding gravity propulsion fields.

4.3.1 The Deflection Effect

The resultant gravity propulsion fields that are directed away from and toward the center of the disk collide at an angle. This results in strong gravity propulsion fields radiating up and down along the center axis. This deflects normal vertical gravity propulsion fields around the disk. This principle was as stated in section 3.2.3 #4, "Gravity propulsion fields are deflected away from increasing gravity propulsion field intensity."

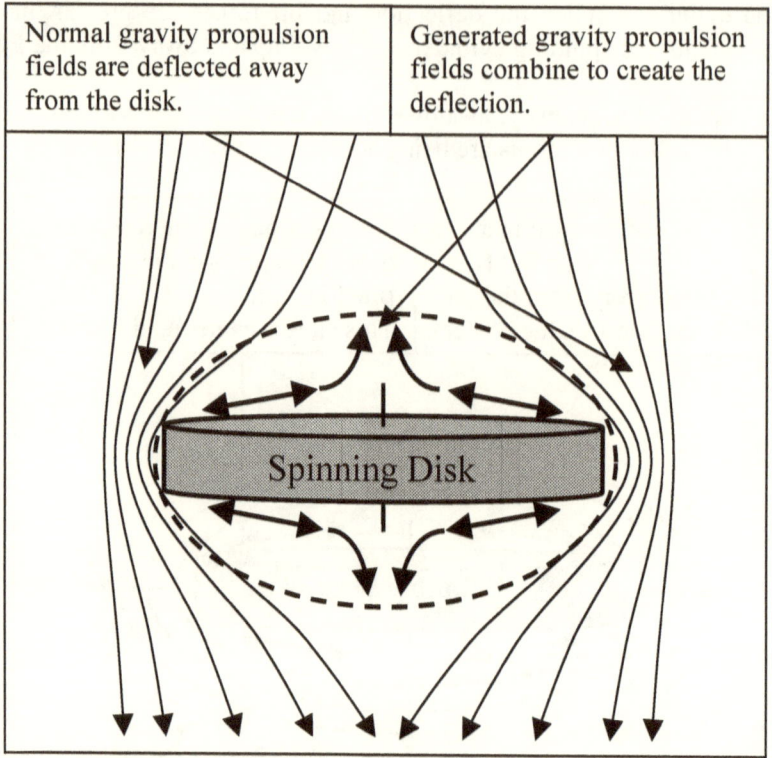

| Normal gravity propulsion fields are deflected away from the disk. | Generated gravity propulsion fields combine to create the deflection. |

FIG. 8. Shows a spinning disk deflecting incoming gravity propulsion fields around it.

This example shows a spinning disk. However, it is more efficient to use a spinning cylinder. For the same about of mass and spin, a cylinder will deflect more of the normal gravity propulsion fields around it. This is because a cylinder's radius is smaller and requires less deflection of fields to reach its edge. The cylinder should be at least 4 times longer than its diameter.

4.3.2 The Lift Effect

The spinning disk gets a lift. This is how it happens. Please refer to Figure 9 below. The spinning disk generates gravity propulsion fields (1) all around it. The generated gravity propulsion fields going downward interfere with the Earth's mass fields that are going upwards (2). This creates gravity propulsion fields going in all directions (3) from the center of the interference. The resulting gravity propulsion fields (3) that travel upward lift the disk.

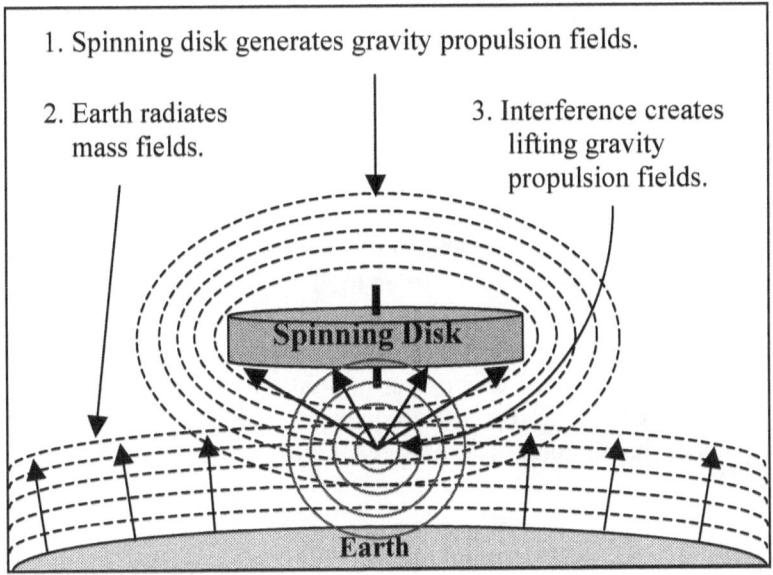

FIG. 9. Shows a spinning disk creating an upward gravity propulsion force by interference of its outward gravity propulsion fields with Earth's mass fields.

4.3.3 The Time Effect

TGP predicts that time speeds up around the edges of a spinning disk. This is because the speed that electrons orbit is proportional to the intensity of gravity propulsion fields. This is explained more in section 6.0 "Time". When the disk is spinning fast enough so its weight is zero, gravity propulsion fields at the edge will be about seven times normal. (This is an estimate.) Time goes little faster there too. As you go outward from there, time and gravity exponentially returns to normal.

Time slows down, directly underneath and above the disk at the center. Gravity is reduced about ⅓ of normal. (This is an estimate.) Time slows down there too. As you go outward from there, time and gravity returns to normal.

4.3.4 The Time Shift

TGP predicts that we should be able to see these unorthodox changes in the speed of time by observing corresponding frequency shifts. This is explained later on, especially in section 7.10.2. TGP predicts the following:

The Time Shift is the apparent variation in frequency due to the difference in the speed of time between the source and the destination.

When light comes from a slower to a faster speed of time, light frequency is shifted lower. When light comes from a faster to a slower speed of time, light frequency is shifted higher.

TGP predicts that as time slows down, frequencies radiated or reflected from the mass are shifted lower than normal. Lower frequencies become redder. As time speeds up, the frequencies radiated from the mass are shifted higher than normal. Higher frequencies become bluer. I call this the time shift.

Light reflected off the spinning disk should show this time shift. Light reflected off the sides of a spinning disk should become bluer and less intense. Light reflected off the top of the disk should become redder and more intense near the center. The visible light range is relatively small from red (4.3×10^{14} hertz) to violet (7.5×10^{14} hertz).

When the cylinder reaches zero weight, TGP estimates that gravity propulsion fields at the cylinder side will be about 7 times normal. This speeds up time by the formula given Section 6. The increase in the speed of time would be $7/P_0 = 7/2,121,637,049 = 3.29934 \times 10^{-9}$. This increases red frequency by 4.3×10^{14} hertz x $3.29934 \times 10^{-9} = 1,418,716$ Hz. This estimate shows that the difference is not large enough to see but can be measured by a spectrometer or interferometer by reflecting a laser beam off the side of the spinning cylinder.

When the cylinder reaches zero weight, TGP estimates that gravity propulsion fields at the center axis of the cylinder will be about ⅓ of normal. This slows down time by the formula given in Section 6. The decrease in the speed of time is $.333/P_0 = .333/2,121,637,049 = 1.57 \times 10^{-10}$. This reduces violet frequency by about 7.5×10^{14} hertz x $1.57 \times 10^{-10} = 117,750$ Hz. This estimate shows that the difference also is not large enough to see but can be measured by a spectrometer or interferometer by reflecting a laser beam off the top of the spinning cylinder.

Any object in these fields of the spinning disk should also have these time shift effects but to a far lesser extent. For example, you could place a white frame around the spinning disk. A spectrometer or interferometer should be able to measure these time shifts and how they change over distance. See the following illustration.

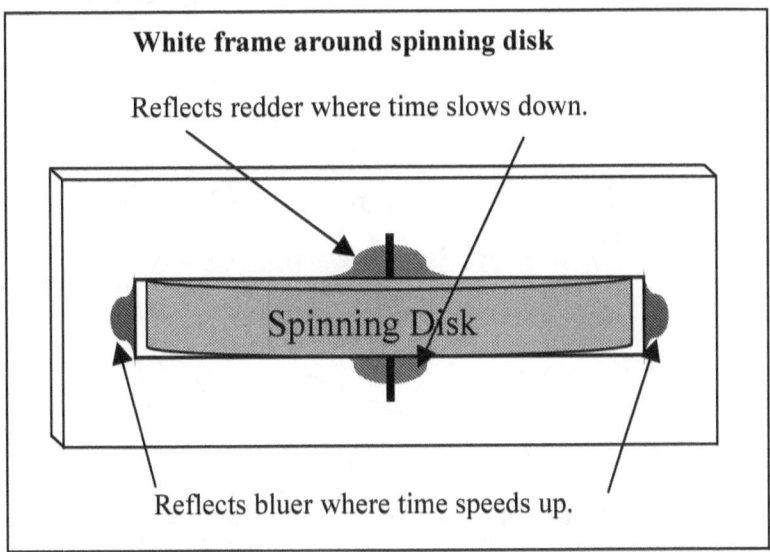

FIG. 10. The time shift effect shows how spinning disk changes gravity propulsion field strength and the speed of time around it.

4.3.5 Spinning Disk Effects (Predictions 1 to 7)

If TGP is correct, we should be able to measure the seven previously predicted effects of a spinning disk or cylinder. It is this type of research that I hope will be of interest to students and professionals in this field. Figure 11 below gives an overview of effects we can measure.

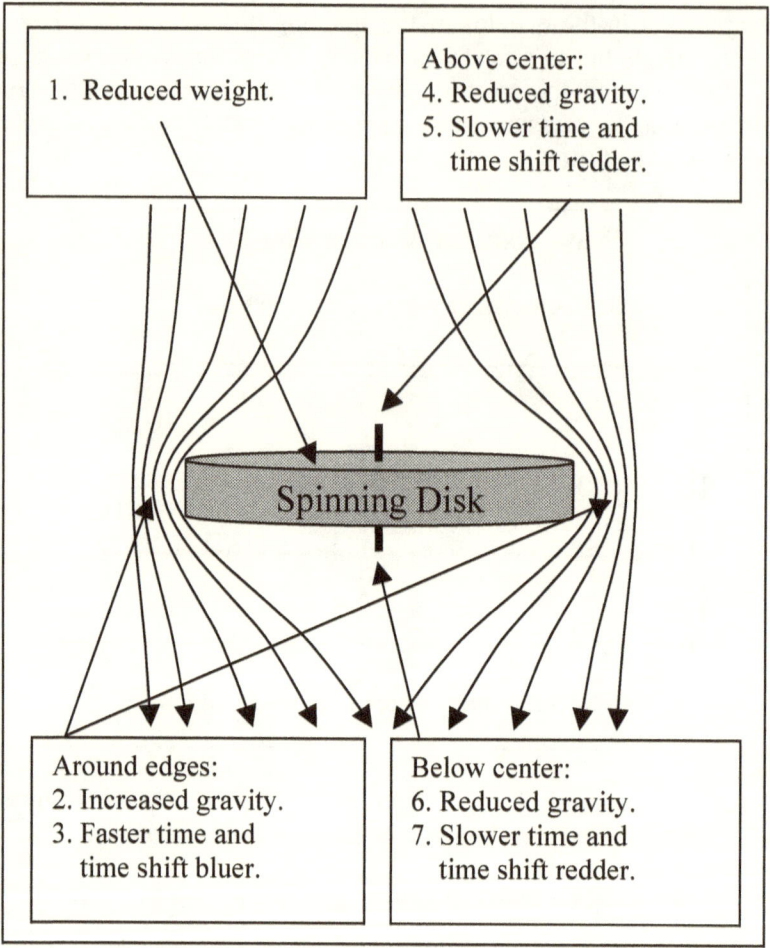

FIG. 11. Seven measurable effects of a spinning disk affecting weight, gravity and the speed of time.

The seven predicted effects vary based upon how fast the disk spins. The effects are proportional to the spin rate of the disk. The faster the disk spins, the greater the effect. These effects are described in more detail below.

1. The disk weight should be reduced. The measuring device should be away from the increased gravity near or under the disk's edges.
2. Near the outside edges of the disk, gravity increases. You can measure this by placing a hoop around the spinning disk. Then measure the weight of the hoop as the disk spins. The gravity effect

exponentially decreases as you move away from the disk. You can measure this by using larger diameter hoops.

3. Near the outside edges of the disk, the speed of time increases. You can measure this by reflecting laser light off the edge. The color will become bluer. This was discussed in the previous section "The Time Shift".

4. Above and near the disk axis, gravity is reduced. You can measure this by hanging a small weight near the center of the disk.

5. Near the disk top axis, the speed of time is slower. You can measure this by reflecting laser light off the top of the disk. You can also measure this by placing a white object near the center. The color will become redder. This was discussed in the previous section, "The Time Shift".

6. Below and near the disk axis, gravity is reduced. This should be the same amount of reduction as in effect #4.

7. Near the disk bottom axis, the speed of time is slower. This should be the same amount of reduction as in effect #5.

To see the best effects, you should spin a cylinder. The cylinder length should be over 4 times its diameter. The faster it spins, the greater its deflected gravity propulsion fields will be. The cylinder needs to be precision spun balanced. Changing the rotation speed (RPM) of the cylinder should proportionally change any measured changes to gravity and time.

The experiment should be done in a vacuum. This eliminates air vortex and refraction as variables in these effects. It is desirable to use a scanning laser to measure frequency shifts from the axis on out. This efficiently shows how fields change per radius for a given speed (RPM).

Warning, if you spin the cylinder too fast it will fly apart. If it is not properly balanced, it could fly outward like a deadly bullet. Lets say you have a 1-inch diameter cylinder made of steel with a tensile strength of 250,000 pounds per square inch and density of .283 pounds per cubic inch. The steel will pull apart if you spin it faster than 31,000 RPM. Material with a higher tensile strength can spin faster. If the cylinder is long enough, it will overcome its weight and lift off. The equation for the maximum speed and radius is:

Tensile Strength = $(4/3)D\Pi^2R^2r^2$

Where:
Tensile Strength of material (lb/in^2 or kg/cm^2)

25

D = Density of material (lb/in^3 or kg/cm^3)
Π = Pie
R = Revolutions per second
r = Radius of cylinder (in or cm)

See Appendix B for my derivation of the equation.

4.4 Spinning Projectiles (Gun Shells)

Spinning projectiles is an early example of the gravity Lift Effect. The idea of spinning an object fast to make them fly better is not new. In 1861, the British created the Armstrong gun that was used on the HMS Warrior. At that time, the HMS Warrior was the most powerful warship in the world. The Armstrong gun had a special improvement. It had a rifled barrel that spun shells when they were shot. The spinning improved the accuracy by providing aerodynamic stability. And, it gave shells an amazing range. It fired 110-pound (50Kg) shells about 2.5 miles (over 4500 yards, 4150 meters). Now we know why. The spinning of shells also gave them a gravity lift. Since then, spinning projectiles quickly became a standard feature in all-large guns.

4.5 Searl's Flying Saucer

John Roy Robert Searl invented the first flying saucer. He calls them "Levity Discs". He discovered the gravity propulsion effect by accident. In 1947 when he was around 14 years old, he tested his first Searl Effect electrical generator. He states it flew off the table and hit the wall. If so, this should be a famous day in history. Since then he has built about 40 more Levity Discs. In 1952 he tested one in an open field in England. He states it rose to 50 feet and stayed there for a while. Then it accelerated at a fantastic speed flying up and out of sight. He never saw it again.

Most of his Levity Discs were used only once. They flew up and away forever. Then he learned how to control them. He built a Levity Disc called Demo 1 that he could control. Demo 1 had many fights.

His Levity Discs fly because they have 66 cylinders rotating at high speeds. All cylinders are solid and rotate around a vertical position. Any cylinder that rotates as fast as Levity Discs do in the vertical position will lift off too. All of these cylinders rotating together deflect incoming gravity propulsion fields around each cylinder and around the Levity Disc. Each cylinder generates its own gravity propulsion fields. These fields combine with the Earth's mass fields. This results in a gravity propulsion lift upwards.

It is the fast spin of the cylinders that provides the lift. Mr. Searl's unique design gets the cylinders to rotate fast. The cylinders are pushed around by a method he invented called the Searl Effect. The cylinders float within an imprinted magnetic field that he invented. This allows them to spin rapidly without mechanical friction.

For more information, see his web site www.searleffect.com or get the book "Antigravity: The Dream Made Reality" by John A. Thomas Jr.

4.6 Podkletnov's Experiment

In 1992, Russian physicist Evgeny Podkletnov published, in the science journal "Physica C", the results of an experiment where he measured the gravity propulsion Deflection Effect.[2] He demonstrated and measured weight reduction in objects above a rotating disk's center. Various materials were used including wood, plastic, ceramic and metal. Weight loss was the same percentage independent of the material used.

He spun a disk that was 25.5 cm (10.8 inch) in diameter and 1 cm (.39 inch) thick. It had a hole in the middle 8 cm (3.15 inch) in diameter. As the disk was spun up to 5,000 RPM, objects above the disk steadily lost weight up to .23% as shown in the following graph.

As the disk decelerated down to 3,300 RPM, weight loss went up to 2.1% at one point.

---

[2] Eugene Podkletnov, "Weak gravitation shielding properties of composite bulk YBCO superconductor below 70K under e.m. field." LANL database number cond-mat/9701074, v. 3, 16 Sep 1997.

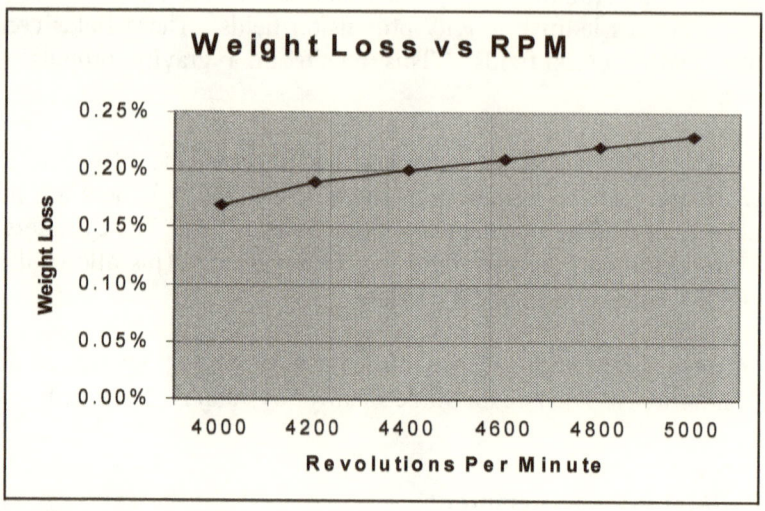

FIG. 12. Graph of weight loss of object above a spinning disk verses RPM of the spinning disk.

The disk was a superconductor. Alternating currents were induced in the disk using a high frequency field. In the above chart, frequency was held at 2 Mhz. With the rotation speed held constant at 4,300 RPM, weight loss increased as frequency increased up to 3.2 Mhz as given in the following graph:

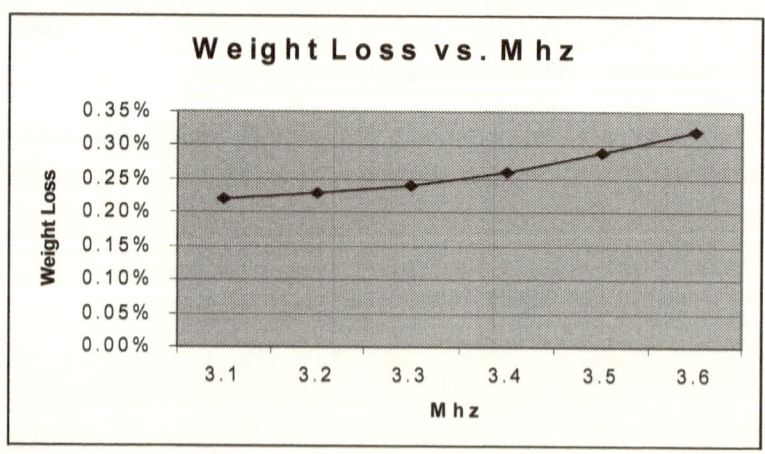

FIG. 13. Graph of weight loss of object above a spinning disk verses induced vibration frequency in disk spinning at 4,300 RPM.

Explanation

Podkletnov's weight reduction was caused by two factors:
1. Rotating the disk.
2. Inducing an alternating current in the disk.

Although, the disk's weight was not measured, it should have lost weight. I estimate that if the disk was a solid cylinder 25.5 cm diameter, 100 cm or more high and spun at 5,000 RPM, its weight would be negative (over 100% weight reduction).

The second reason for weight loss was due to inducing an alternating current in the disk. The alternating current accelerated atoms in the superconductor by vibrating them. The higher the frequency the more efficiently the frequency vibrated atoms inside the disk's super-conducting lattice structure. This created more gravity propulsion fields.

John H. Schnurer, director of applied sciences at Physics Engineering at Antioch College in Yellow Springs, Ohio, duplicated this experiment and got similar results.[3] He spun a solid 1-inch diameter super-conducting disk and achieved a 5% weight loss in a plastic sample over the center of the disk.

4.7 Questions

The answers are in Section 16.

4.1 According to TGP, name the four effects produced by a fast spinning disk.
4.2 According to TGP, on the outside of a spinning disk, does time stay the same, slow down or speed up?
4.3 According to TGP, at the center of a spinning disk, does time stay the same, slow down or speed up?
4.4 Who invented the first flying saucer?
4.5 What was the main thing that Podkletnov's experiment measured?

---

[3] Otis Port, "Antigravity? Well, It's all Up in the Air" (p97), Feb. 17, 1997

5.0 Gravity

In the next few sections, we will explain how gravity works. You can call this the Gravity Field Theory. It explains how gravity is an interaction among mass fields and gravity propulsion fields. Gravity works similar to how inertia resistance works. Compare this section to section 4.1.1. Here is an overview of how gravity works:

Step 1: Mass fields. Remember, "All mass has a mass field around it. Mass fields are energy that radiates out from all particles." (Section 3.1.1) All objects in the universe radiate a mass field around them. This includes stars and planets.

Step 2: Creation of gravity propulsion fields. Also remember, "When a mass field meets another mass field, it creates gravity propulsion fields." (Section 3.1.3) Mass fields radiated from all the objects in the universe combine to produce gravity propulsion fields. Gravity propulsion fields radiate out in all directions. This results in gravity propulsion fields filling the universe. Gravity propulsion fields are all around us coming from all directions.

Step 3: Gravity is the resultant force. "Mass fields push atoms in a mass outward less than incoming gravity propulsion fields push inward on the same mass. The resultant force is what we call gravity." (Section 3.1.3) Here is an example of a person on the earth's surface. The earth's radiated mass fields push the person upward. Gravity propulsion fields push the same person downward. The gravity propulsion fields push the person downward more than the earth's mass fields push the person upwards.

The Gravity Field Equation is:
Gravity = Gravity Propulsion Fields – Mass Fields

We will now explain the above three steps in more detail in sections 5.1 to 5.5.

5.1 Earth and Sun Create Gravity Propulsion Fields

The Earth and Sun radiate mass fields around them. Somewhere in between the Earth and the Sun, their mass fields collide. The interference between the two mass fields results in gravity propulsion fields radiating outward in all directions. In the following illustration, we replace our original masses (from section 3.2) with the Sun and Earth.

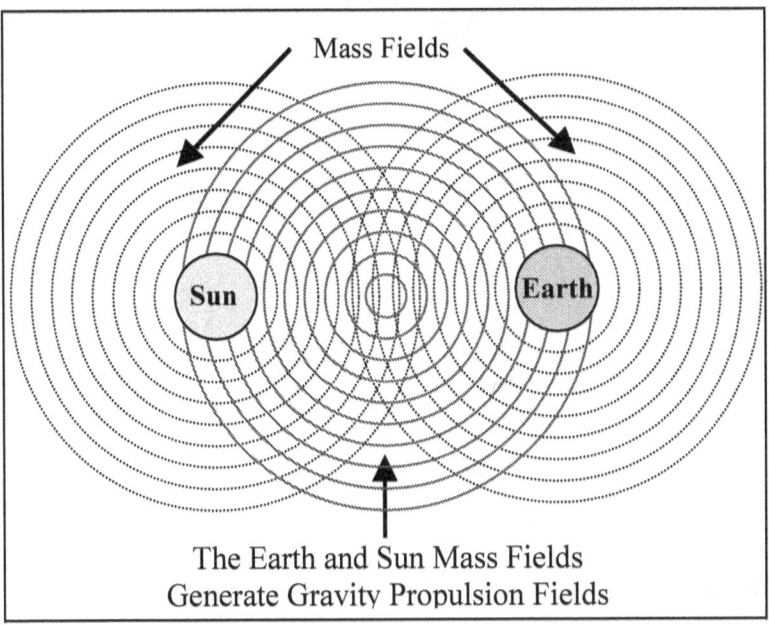

FIG. 14. Sun and Earth mass fields combine to create gravity propulsion fields.

5.2 Gravity Propulsion is Every Where

We can now imagine that all space is filled with gravity propulsion fields that radiate in all directions. All matter radiates mass fields. All stars, planets, asteroids, comets and even space gas radiate mass fields. When these mass fields meet, they interfere with each other and create what we call gravity propulsion fields. Stars, galaxies and planets create gravity propulsion fields. These gravity propulsion fields surround us. Most of the gravity propulsion fields that affect us come from distant stars and galaxies.

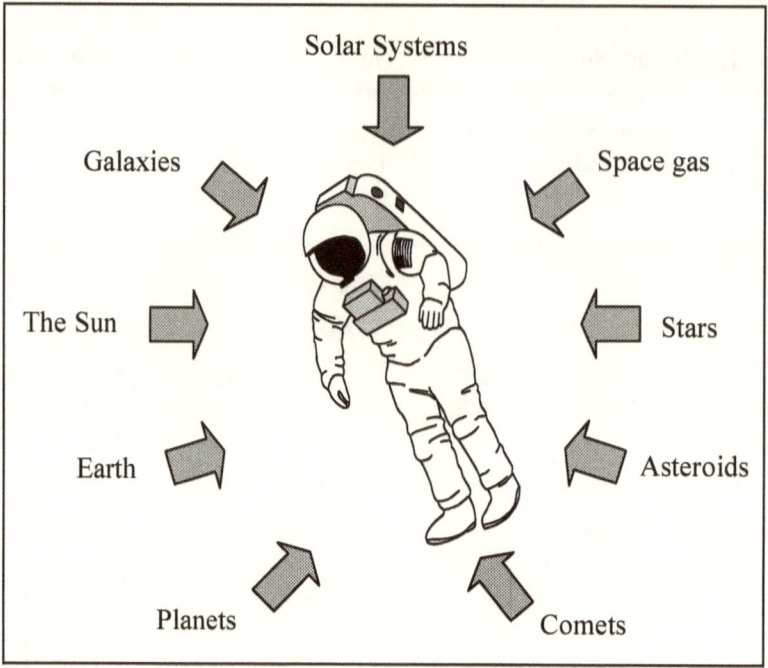

FIG. 15. Gravity propulsion fields come from everywhere and surround us.

5.3 Gravity Defined

According to TGP, here are some definitions of what gravity is:

"Gravity is the resultant force between a mass field and gravity propulsion fields acting on an object."

"Gravity between two objects is the amount of gravity propulsion force that each object absorbs that otherwise would have been absorbed by the other object."

Gravity is a result of shielding that matter does to gravity propulsion fields. When a nucleus absorbs a gravity propulsion field, it shields part of the force from affecting other objects. When an object, like the Earth, absorbs gravity propulsion fields, it shields some of the force from affecting other objects, like you. The Earth under you is shielding you from gravity propulsive fields that would otherwise push you upward. Gravity propulsion is pushing on your body from all **other** directions. The result is that you are being pushed toward the center of the Earth.

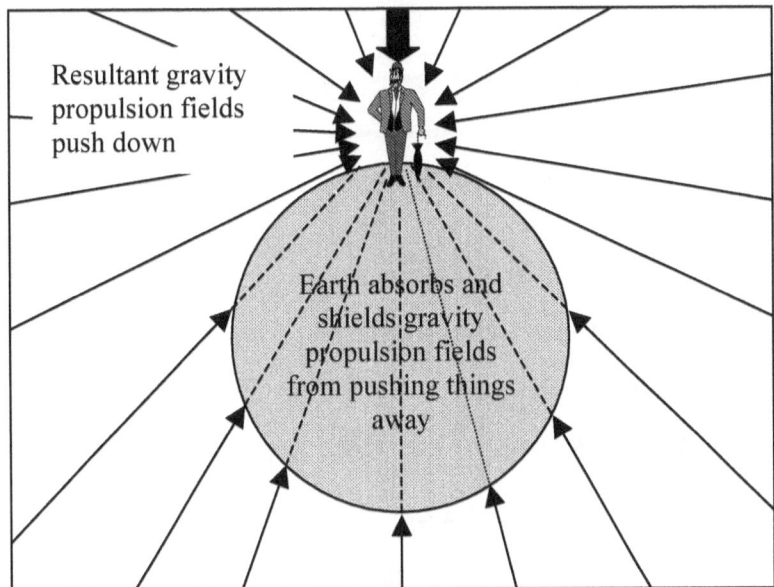

FIG. 16. Resultant gravity propulsion forces push objects towards the earth.

The greater the mass of an object is, the more it shields (stops) gravity propulsion fields from going through it. The resultant gravity propulsion forces pushing down on other objects are what create the gravity effect. Later, in section 9.3 "TGP's Quantum Gravity Theory", we will explain how matter "shields" gravity propulsion forces by aligning quantum strings.

5.4 Gravity Is Like An Upside Down Airplane

Gravity works similar to how an airplane flies. Airplanes don't fly because they are attracted to outer space, even though it looks that way. Neither is an object attracted to the Earth, even though it looks that way. Airplanes fly because the air pressure below the wing, pushing upward, is normal and air pressure above the wing, pushing downward, is less than normal. The result is that the airplane is lifted up because of the difference in the opposing forces. The same principle applies to gravity. It is the difference in opposing forces that cause gravity. With gravity, the resultant forces push downward. With airplanes, the resultant forces push upwards. In this respect, gravity is like an upside down airplane.

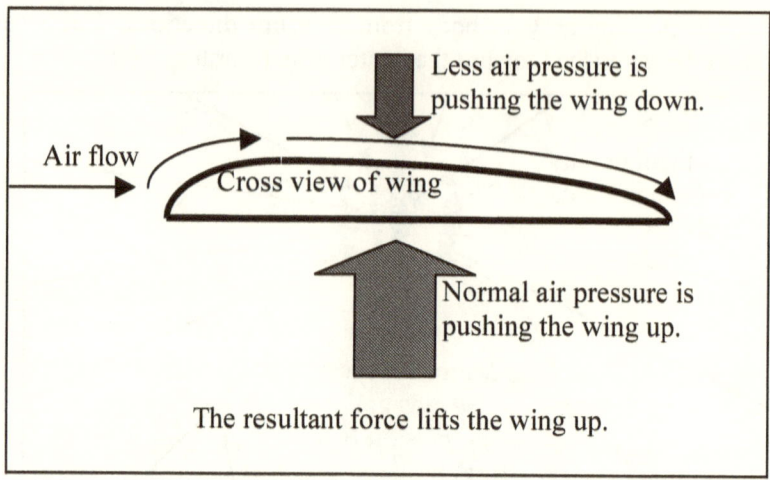

FIG. 17. Resultant air pressure forces push the wing upward.

5.5 Gravity Between Sun and Earth

Gravity propulsion forces are pushing the Earth and Sun together. The Sun shields some gravity propulsive fields from reaching the Earth. The Earth is still being pushed in all other directions. Because there is less gravity propulsion force on the side of the Earth that faces the Sun, the resultant propulsive forces push the Earth toward the Sun.

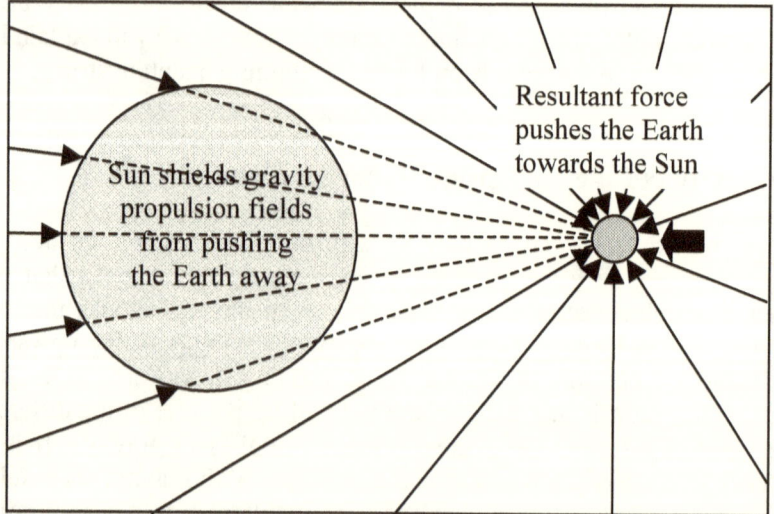

FIG. 18. The Sun's shielding of gravity propulsion fields results in other gravity propulsion fields pushing the Earth towards the Sun.

Also, the Earth shields some gravity propulsive fields from reaching the Sun. The Sun is being pushed in all other directions too. Because there is less gravity propulsive force on the Sun's side that faces the Earth, the resultant gravity propulsion force pushes the Sun toward the Earth (just a little).

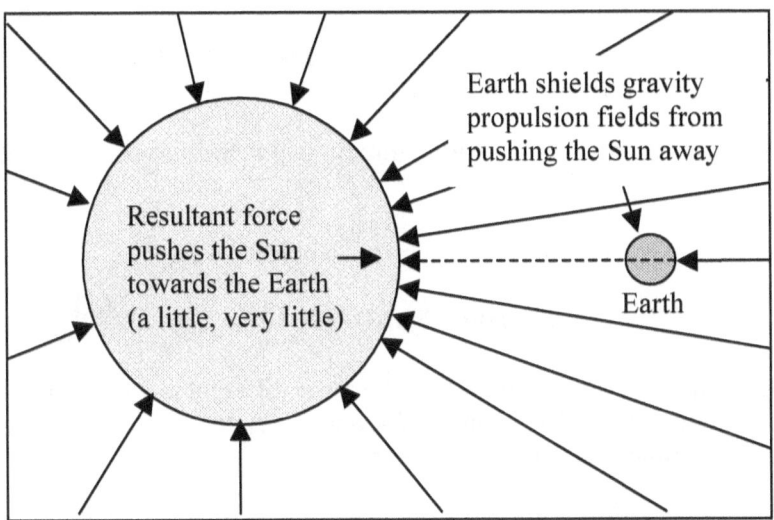

FIG. 19. The Earth's shielding of gravity propulsion fields results in other gravity propulsion fields pushing the Sun towards the Earth.

Gravity between the Earth and the Sun is the resultant gravity propulsive force between them. The gravity between the Earth and the Sun is the total amount of gravity propulsion fields that the Sun absorbs that otherwise would have been absorbed by the Earth PLUS the amount of gravity propulsion fields that the Earth absorbs that otherwise would have been absorbed by the Sun.

5.6 Gravity Equation

TGP presents the delta gravity equation. The general equation below shows the three gravity variables that change the force of gravity as a mass accelerates or decelerates.

$$\Delta G = D_t - D_a - I$$

Where:

ΔG = Change of gravity.

D_t = Downward gravity propulsion fields deflected towards a mass.

D_a = Downward gravity propulsion fields that were deflected away from a mass.

I = Upward gravity propulsion fields on an area that were created by interference of the mass' generated gravity propulsion fields with earth's mass fields.

All units are in units of earth gravity = 9.8 meters/second^2 = 32.2 feet/second^2

The general equation for the change of an object's weight is:

$$\Delta \textbf{Weight} = \sum \textbf{M(D}_t \textbf{ - D}_a \textbf{ - I)}$$

The intensity of gravity propulsion forces is different at various areas of an accelerating mass. The change of a mass' weight is the sum of these vertical forces times the mass in each area.

We can increase, reduce and reverse the force of gravity on an object. In the gravity equation, variables D_t, D_a and I are created by acceleration. Therefore, we can control gravity by accelerating (A) matter. We can accelerate matter three ways:

1. Accelerate it in a straight line.
2. Rotate it.
3. Vibrate it.

An exciting prediction of TGP is, "If we rotate or vibrate matter fast enough, it will overcome the force of gravity and lift up." All three variables D_t, D_a and "I" can contribute to reduce and reverse an object's weight.

1. The variable D_t represents gravity propulsion fields that are deflected downward towards an area. With enough acceleration, most of these downward fields can be deflected away from an object, thereby reducing its net weight.

2. The variable D_a represents gravity propulsion fields that are deflected away from an area. This contributes to reducing an object's weight.
3. The variable "I" represents increased upward gravity propulsion fields. This contributes to reducing the object's weight and lifting it.

The following two examples show how these three variables apply to a spinning object.

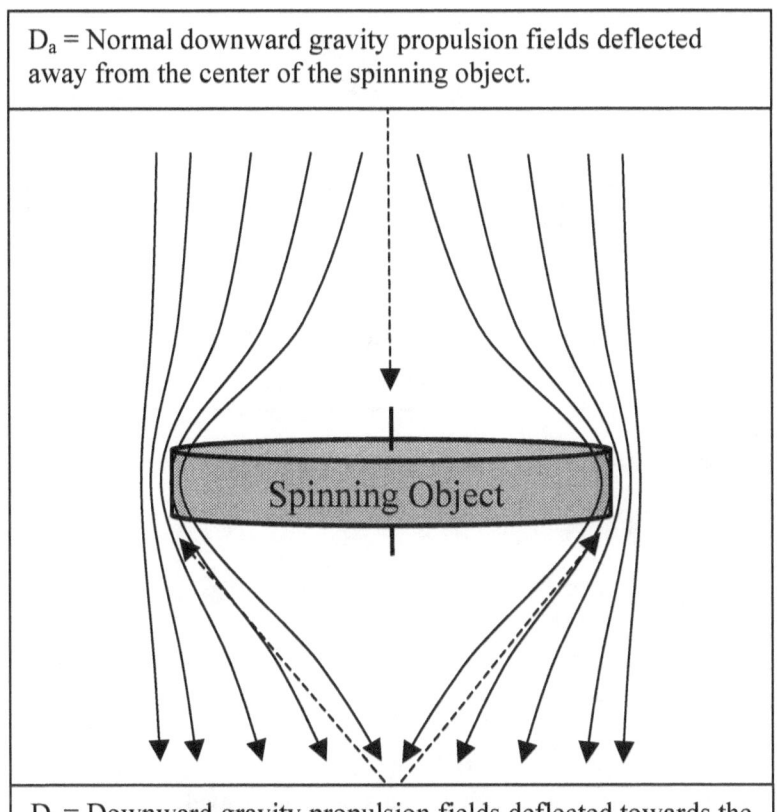

D_a = Normal downward gravity propulsion fields deflected away from the center of the spinning object.

Spinning Object

D_t = Downward gravity propulsion fields deflected towards the outside of the spinning object. Note that some deflected fields miss the object.

FIG. 20. Shows a spinning object creating deflected gravity propulsion fields D_a and D_t.

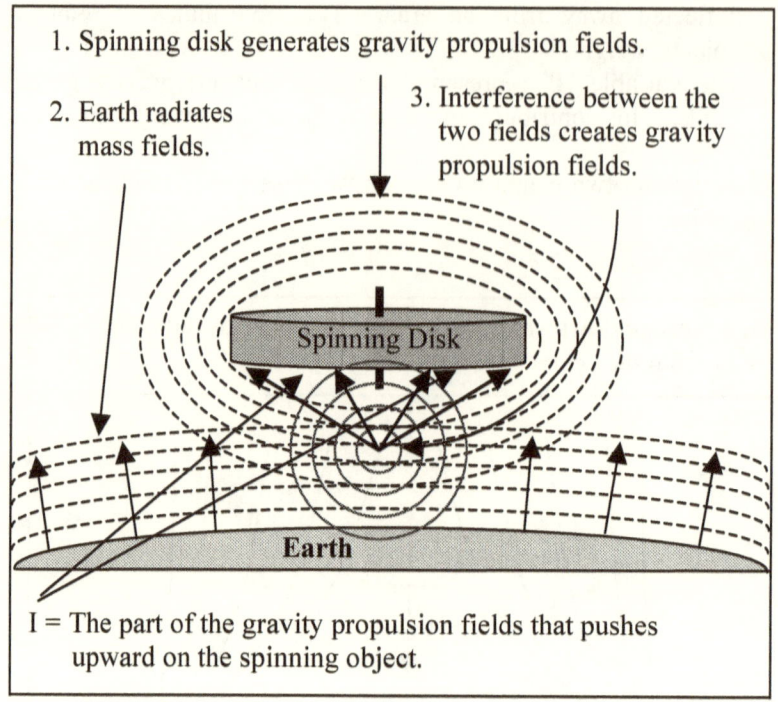

1. Spinning disk generates gravity propulsion fields.

2. Earth radiates mass fields.

3. Interference between the two fields creates gravity propulsion fields.

Spinning Disk

Earth

I = The part of the gravity propulsion fields that pushes upward on the spinning object.

FIG. 21. Shows a spinning object creating upward gravity propulsion field "I".

5.7 Questions

The following questions are about TGP as stated in this section. The answers are in Section 16.

5.1 What direction do mass fields travel?
5.2 What direction do gravity propulsion fields travel?
5.3 Gravity is the resultant force between what two fields?
5.4 When the earth's mass field and sun's mass field meet, what do they create?
5.5 What are the gravity variables that can contribute to reduce an object's weight?

6.0 Time

Time is a way that we measure a succession of actions or events. We currently measure time by the atomic clock. The atomic clock is based on the orbits and energy states of electrons. Since 1967, the International System of Units (SI) has defined the second as the period equal to 9,192,631,770 cycles of the radiation that corresponds to the transition between two energy levels of the ground state of the Cesium-133 atom. This definition makes the cesium oscillator the primary standard for time and frequency measurements.[4]

The atomic clock may give us the illusion that time is a constant that we can depend up. But time is not a constant. Time is not a variable. Time is not a force. Time does not and cannot make anything happen. Time is an illusion. Time is the result of all forces that make things happen. Change these forces and you change time. TGP gives us the understanding that we can change time. The following sections explain the forces that can change the speed of time and derives the general equation for time.

6.1 Developing The Time Equation

In the next few sections, we are going to develop the time equation. There is some math involved but its not complicated. Time is directly proportional to gravity propulsion forces. The stronger gravity propulsion forces are, the faster time goes. The weaker gravity propulsions forces are, the slower time goes. We can represent this by the following equation:

Time = Gravity Propulsion
Time = P

The letter "P" represents the total amount of gravity propulsion force on an object or area in these time equations.

Electrostatic forces are proportional to the intensity of gravity propulsion fields around it. The stronger gravity propulsion forces are, the faster electrons orbit around the nucleus and the faster time goes. The weaker gravity propulsion forces are, the longer it takes electrons to orbit around the nucleus and the slower times goes. When gravity propulsion

---

[4] www.boulder.nist.gov/timefreq/general/precision.htm, 1/23/02

forces are zero, electrons stop orbiting and time stops. In other words, when gravity propulsion = 0 then time = 0.

There are four forces that can change the intensity of gravity propulsion forces (P). These forces are:
1. Gravity
2. Acceleration
3. Deflections
4. Interference

These forces that can change the normal gravity propulsion force are represented this way:

Time = P + ΔForce

To make gravity measurements (a force) equal time measurements (seconds), we will divide the right side by P. This normalizes the equation to be relative to time. The normalized time equation is:

Time = 1 + ΔForce/P

When Time = 1, this means that time goes at normal speed when total gravity propulsion forces equal P. Lets say there is a Force that increases gravity propulsion and makes Time = 1.01. This means that time sped up by 1%. Lets say there is a Force that decreases gravity propulsion and makes Time = .99. This means that time slowed down by 1%.

6.2 Gravitational Time Shift

This section describes how gravity changes time. Gravity (G) is a force that directly reduces the gravity propulsion force. Therefore gravity reduces the speed of time. The time equation for this is:

Time = 1 – G/P

The amount that gravity (G) reduces the total amount of gravity propulsion force (P) is relative to the total amount of gravity propulsion force (P). This is why we divide G by P. It is this percentage that the speed of time is reduced by.

See the following Figure 22 for this example. Image that the force on "P" as a cube in space with gravity propulsion forces pushing equally on all six sides. Now we put a mass under side 1 of the cube. Side 1 has some gravity propulsion fields blocked by the mass. The mass' gravity reduces some of the gravity propulsion forces pushing on P from side 1. The reduction of gravity propulsion slows time.

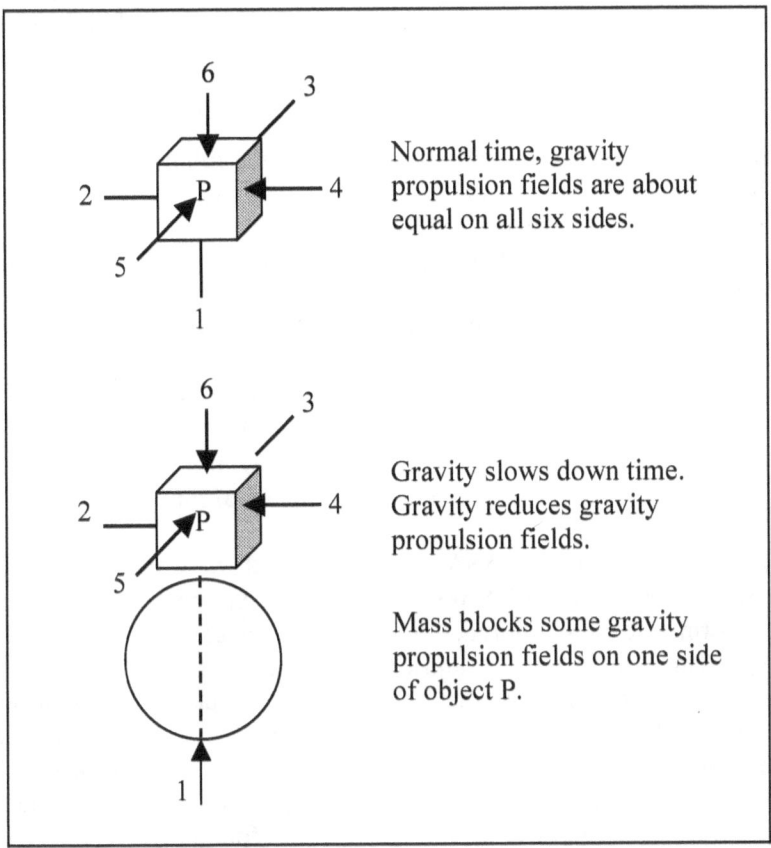

Normal time, gravity propulsion fields are about equal on all six sides.

Gravity slows down time. Gravity reduces gravity propulsion fields.

Mass blocks some gravity propulsion fields on one side of object P.

FIG. 22. Gravity reduces total gravity propulsion forces on point P.

6.3 The Force. How Strong Is It?

In this section, we will calculate the strength of the gravity propulsion force. The gravity propulsion force (P) around us is about 2,121,637,049g. This means the gravity propulsion force is over 2 billion times more powerful that of earth's gravity (g). Gravity propulsion (P) is measured in units of earth's gravity: 9.8 meters/second^2 or 32.2 feet/second^2.

This is the force that holds matter together and keeps electrons spinning. We don't notice it because the force is almost equal on all sides. This is like the air pressure around you. The pressure of air around you at sea level is about 14.7 pounds per square inch (1 kg per square centimeter). We normally don't notice the air pressure because it is equal on all sides. However, there is a slight difference in the gravity propulsion force around us that we notice. We call this difference gravity.

From the above time equation we can derive the following:
Time = 1 – G/P
Change of Time = -Change of Gravity / P
                (Look only at what changes.)
P * Change of Time = -Change of Gravity
                (Multiplied both sides by P.)
P = -Change of Gravity / Change of Time
                (Divided both sides by Change of Time.)

The negative sign means as gravity decreases, the speed of time goes faster. It also means as gravity increases, the speed of time slows down.

Thanks to the work at NASA together with others, we can calculate the value of P. In 1976, NASA/SAO completed an experiment to measure the change of time at a high altitude. The project was called Gravity Probe A (GP-A).[5] This was to test Einstein's general theory of relativity and verify his gravitational red shift prediction. They developed an accurate clock called the Hydrogen Maser Clock (HMC). The clock frequency had an accuracy of 1 part in 10^{16}. A Scout D rocket launched the clock in a two-hour sub-orbital flight. The rocket coasted up to 10,000 km altitude and then fell back to Earth. The result was that they found a frequency shift of (f/f -1) = $4*10^{-10}$ at 10,000 km altitude. This means that the speed of time increased by $4*10^{-10}$, which is $4*10^{-8}$ percent faster. The experiment verified Einstein's predicted Gravitational Red Shift with an accuracy of 70 parts per million.

---

[5] "Tests of relativistic gravitation with a spaceborne hydrogen maser", R.F.C. Vessot, M.W. Levine, E.M. Mattison, E.L. Blomberg, T.E. Hoffman, G.U. Nystrom, B.F. Farrell, R. Decher P.B. Eby, C.R. Baugher, J.W. Watts, D.L. Teuber and F.D. Wills, Physical Review Letters, Vol. 45, Dec. 1980, pp. 20812084.

Here is how we calculate the gravity propulsion force:

1. Radius of the earth is 6,378,137 m at the equator and 6,356,752.3142 m at the poles.
2. Average earth radius; r = (6,378,137 + 6,356,752.3142) / 2 = 6,367,444.657 m
    I assume this is the earth's radius where the rocket was launched.
3. 10,000 km altitude in terms of earth radius from earth's center = 10,000,000/6,367,444.657 + 1 = 2.57048871855518 r
4. Gravity at 10,000 km = $1/r^2$ = $1/2.57048871855518^2$ = 0.151345180507558
5. Change of gravity at 10,000 km = 0.151345180507558 - 1 = -0.848654819492442. This means gravity is reduced by -84.8654819492442% at the 10,000 Km altitude.
6. Change of time at 10,000 km = $4*10^{-10}$. This is from the above experiment.
7. P = -Change of Gravity / Change of time. Equation on previous page.
8. P = $0.848654819492442g / 4*10^{-10}$. Values for #7.
9. P = 2,121,637,049 times earth gravity.

P is in units of earth's gravity at earth's radius of 6,367,444.657 meters. Earth's gravity is approximately 9.8 m/s^2 or 32.2 ft/s^2. More accurate information of the earth's radius and gravity at the location the rocket was launched and the resulting frequency shift will increase the accuracy of P.

The time equation is now this:

Time = 1 – (G / 2,121,637,049g)

The gravity propulsion force pushing on you is equivalent to the energy of about 500 tons of TNT exploding every second. The gravity propulsion force is not destructive because it is evenly distributed around you. It is this force that holds your atoms together.

The following describes how this is calculated. The gravity propulsion force that is pushing on you, is about 2,121,637,049 (over 2 billion) times earth gravity. When we say "times earth gravity" this means times 9.8 meters/second^2 or 32.2 feet/second^2. P is a force on a mass according to F = MA. The P force in Newtons = Mass x P in the metric system.

The gravity propulsion force, measured in Newtons, is:

= Mass x P x 9.8 m/s^2.

= Mass x 2,121,637,049 x 9.8

= Mass x 20,792,043,080

Where Mass is in kg. A Newton is a force that accelerates a 1 kg mass at the rate of 1 meter per second^2.

Let's say a person weighs 100 kg (about 220 pounds). The total gravity propulsion force on them would be 2,079,204,308,000 Newtons.

Ton of TNT = 4.184×10^9 Newton-meters (Joules)

2,079,204,308,000 / 4.184×10^9 = 497 tons of TNT per second.

Therefore, the gravity propulsion force pushing on you is equivalent to about 500 tons of TNT exploding every second. It is this force that is holding your body together.

6.4 Gravity Propulsion is Relative (P$_0$)

Within our solar system, gravity propulsion forces slightly vary. Gravity reduces gravity propulsion forces. Relative to earth, the top four gravities that affect us are listed below with the strongest listed first:

1. Earth
2. Moon
3. Sun
4. Jupiter

When you measure gravity propulsion forces in space it's relative to the gravities around you. The gravity propulsion force calculated above is relative to earth's orbit around the sun. When you move closer or further away from the sun, the gravity propulsion forces change. It is also relative to how far away the moon is.

Because the value of P changes, we will use P$_0$ ("o" is pronounced naught) to represent gravity propulsion forces in space at earth's orbit around the sun when the earth's gravity is not present. The time equation is now represented as:

Time = 1 – G/P₀

Where P_0 = 2,121,637,049. When Time = 1, this is the speed of time along earth's orbit when earth's gravity is not affecting it. Time on earth is slower because of earth's gravity. According to the time formula, when G = 1, the speed of time on earth is:

Earth time = 1 – 1/2,121,637,049

This means that Earth time is 0.999999999528666 slower than the speed of time in space without earth's gravity.

6.5 Deflection Time Shift

Deflections decrease gravity propulsion fields in some areas and increase them in other areas. For example, centrifugal force causes deflections. As an object spins, it deflects gravity propulsion fields from the center of the object on out, which reduces the speed of time. Gravity propulsion fields that are deflected away from an area increase gravity propulsion fields another area, which increases the speed of time. We can control the speed of rotating objects and thereby control deflection of gravity propulsion fields. This means we can increase the speed of time in some areas and reduce the speed of time in other areas. The reasons for these deflections are explained in more detail in the Quantum String Theory.

"D_a" represents gravity propulsion fields that are deflected away from an area. "D_t" represents the gravity propulsion fields that are deflected towards an area. This adds D_a and D_t to the time equation, which now is:

Time = 1 + (D_t – D_a – G)/P₀

Having both D_t and D_a in the time equation, makes it a general equation. It can now account for objects that accelerate in a straight line, circle or curve.

6.6 Interference Time Shift

Acceleration creates gravity propulsion fields radiating from the object. These fields interfere with the earth's mass field and become stronger gravity propulsion fields. Where these gravity propulsion fields meet the mass, it adds to the gravity propulsion fields affecting it. The variable "I" represents the increase of gravity propulsion fields due to gravity field interference. This adds "I" to the time equation. The time equation now is:

$$\text{Time} = 1 + (D_t - D_a + I - G)/P_o$$

6.7 The Time Equation

The final general time equation is:

$$\text{Time} = 1 + (D_t - D_a + I - G)/P_o$$

Where:
Time = 1 = Speed of time in earth's orbit when unaffected by earth's gravity.
G = 1 = Earth's gravity at a radius of 6,367,445 km.
D_t = Gravity propulsion fields deflected toward the mass.
D_a = Gravity propulsion fields deflected away from the mass.
I = Interference among mass fields and gravity propulsion fields that increase gravity propulsion fields in the mass.

P_o = 2,121,637,049 times earth gravity.
All units are in units of earth gravity = 9.8 meters/second^2 = 32.2 feet/second^2

6.8 Questions

The following questions are about TGP as stated in this section. The answers are in Section 16.

6.1 What causes time?
6.2 When gravity propulsion fields are stronger, how does it affect time?
6.3 How does gravity affect time?
6.4 How does interference of mass fields with generated gravity propulsion fields affect time?
6.5 Is it possible for people to increase and decrease the speed of time?

7.0 Wonders of the Universe, Part 1

Based on our new understanding of gravity fields, we can now explain the following wonders of the universe:
1. What created the universe's big bang.
2. Why the universe is expanding and why its expansion is accelerating.
3. The speed of time changes throughout the universe.
4. The speed of light changes throughout the universe.
5. Our speed of light is speeding up.
6. Our time is slowing down.
7. Matter is leaving us forever.
8. The universe will end.

After we explain TGP's Quantum String Theory, we will explain more wonders of the universe (part 2) in section 10.

7.1 The Big Bang

TGP is compatible with the "Big Bang Theory". The Big Bang Theory explains that the universe began with a big bang that created all mass and energy. The universe has been expanding outward ever since. The Big Bang Theory helps to explain the role of gravity propulsion. TGP predicts, "Gravity propulsion is the pressure inside this huge Big Bang cosmic explosion."

The Big Bang's gravity propulsion forces are similar to the pressure inside of a bomb exploding in space. The table below shows this comparison.

The Big Bang	A Bomb Exploding In Space
In the beginning, there were tremendous gravity propulsion forces in the center.	In the beginning, there was a tremendous pressure in the center.
Gravity propulsion fields and forces bounced back and forth among the elements that are rapidly expanding outward in all directions.	The pressure is particles that bounce back and forth among the elements that are rapidly expanding outward in all directions.

47

As the outer elements rapidly expand into the vacuum of space, the gravity propulsion forces quickly disperse.	As the outer elements rapidly expand into the vacuum of space, the pressure and energy quickly disperse.

We are living in the mist of this huge explosion which is still going on. TGP predicts that we have been losing useful energy and mass outward ever since.

7.2 Expansion of the Universe is Accelerating

TGP predicts the expansion of the universe will constantly accelerate. This is because gravity propulsion (GP) is pushing all matter away from the center of the universe. GP is accelerating the expansion of the universe. Galaxies that are furthest away from the center of the universe are accelerating the fastest. This is because there is no outside force to slow them down. There is just the vacuum of space on the outside and gravity propulsion from the inside accelerating them outward.

Recent evidence confirms that the expansion of the universe is accelerating. In January 1998, the Supernova Cosmology Project, based at Berkeley Lab, presented the first compelling evidence that the universe expansion is accelerating.[6] Soon after that, the High-z Supernova Search Team announced that the data they collected also supports that the universe expansion is accelerating. Since then there have been more discoveries confirming that the universe expansion is accelerating.

TGP is the only theory that explains why the universe expansion is accelerating.

7.3 Cosmic Gravity Propulsion (P_c)

TGP further predicts that the intensity of gravity propulsion forces vary throughout the universe. The center of the universe has the strongest gravity propulsion forces. This is where time goes the fastest. There, electrons orbit atoms very fast. The center of the universe has the strongest gravity

---

[6] "Science Magazine Names Supernova Cosmology Project 'Breakthrough of the Year'", Berkeley Lab Research News, 12/17/1998. Also see www.lbl.gov/supernova.

propulsion forces because the matter around it keeps most the GP forces from escaping. It is matter that radiates gravity propulsion forces back inward.

Near the edges of the universe gravity propulsion forces are the weakest. This is where time goes the slowest. Electrons there orbit atoms very slowly in large circles.

Within our own solar system, we will not be able to measure the difference of this cosmic gravity. Cosmic gravity is evenly distributed throughout our solar system. Our solar system is only about 7.3 billion miles in diameter. Compared to the size of the universe, our solar system is just a microscopic spec on a large map. In our theory, one side of our solar system is going slower than the other side. But the difference is probably too small to measure.

Cosmic gravity propulsion is pushing and is accelerating our solar system towards the outer edge of the universe.

7.4 Time is Slowing Down

As our universe expands, gravity propulsion forces become weaker. Our speed of time slows down. Atom's mass is reduced. Our gravity gets weaker. Electron orbits get larger. The rate that electrons orbit around the nucleus takes longer. Atom's reactions to other atoms slow down. As time goes on and orbit rates slow down, light travels further for the same number of electron orbits. Basically, time slows down. But we don't notice it because everything looks the same. It's all relative.

7.5 The Speed of Light is Speeding Up (Prediction 8)

TGP predicts that our speed of light is speeding up. We can measure this. Actually, the speed of light around us stays about the same. It is our speed of time that is slowing down. From our point of view, the speed of light appears to be speeding up. We can measure how much our time is slowing down by how fast the speed of light is speeding up. We can do this by measuring the meter again. We have to compare a light meter length against the *physical* meter length. We should compare the 1960 light meter against the *physical* 1960 International Prototype Meter. This allows us to see how fast the speed of light has increased in over 40 years. Also note that the physical meter length increases too as electron orbits get larger but it does not increase as much as the speed of light increases.

On October 14, 1960, the Eleventh General Conference on Weights and Measures redefined the International Standard of Length of the meter as 1,650,763.73 vacuum wavelengths of light resulting from unperturbed atomic energy level transition 2p10 - 5d5 of the krypton isotope having an atomic weight of 86. The wavelength is λ = 1m/1,650,763.73 = 0.605,780,211 μm[7]. I call this the 1960 light meter for short.

The 1960 light meter was the exact length of Prototype Meter No.27. Prototype Meter No.27 is made of a platinum-iridium alloy. It has a X-shaped cross section. The meter was defined as the distance between two graduation lines in the bar at 0 °C. If we compare the length of this physical meter against the 1960 light meter today, we should find that the physical meter has shrunk in size. Another way to say the same thing is that the speed of light is now faster.

Here is the history of Prototype Meter No.27. On May 20, 1875, twenty countries signed the Treaty of the Meter, including the United States, at the International Metric Convention. This established the International Bureau of Weights and Measures (Bureau Intérnational des Poids et Mésures, BIPM). In 1889, the BIPM created the International Prototype Meter. The BIPM made 40 copies of the International Prototype Meter and gave two copies to each member country. All meter bar calibrations were done by comparisons in optical comparators. Prototype Meter No.27 served as the U.S. primary standard of measurement from 1889 to 1960. This meter is now on exhibit in the NIST Museum at Gaithersburg, Maryland. This is just one of the 40 copies of the International Prototype Meter.

Someday, our time will be going so slow causing light to travel so fast that it will be difficult to comprehend. I think it has already happened. Light currently travels at about 186,000 miles a second. That's 669,600,000 miles an hour! Can you comprehend that speed? I can't. We are going very, very slow indeed. And we are going slower every moment. It wasn't always like that. At one time in the past, light was probably traveling at only 4 times the speed of sound. The speed of sound is about 760 mph at sea level. So the speed of light then would be 3,000 miles an hour. Now we are going about 223,200 times slower than that. In one second of our

---

[7] www.mel.nist.gov/div821/museum/timeline.htm, 1/20/02

current time, over 2 and a half days happened back then. Within 4 hours, of our current time, over a hundred years would have happened then.

7.6 The Mass Exodus

TGP predicts that we are constantly losing mass. Some outward-bound gravity propulsion fields leave us forever. This is because they never hit any matter on their way out and therefore never have a chance to return inward again. Some of the outward-bound gravity propulsion fields hit some matter accelerating the matter outward. These fields are then recycled. They become radiated mass fields. Some of these fields are radiated back inward. It is these few remaining gravity propulsion fields, which return inward (bounce back) that end up pushing matter together. Then, a moment later, these fields are radiated back out. Some gravity propulsion fields head for the edge of the universe never to return. This process happens near the speed of light. Throughout the universe, gravity propulsion forces are getting weaker.

As gravity propulsion leaves us and the remaining forces get weaker, matter is converted to energy. Here is how it happens. As atoms radiate their mass field, their mass is reduced by the amount of energy they radiate. $E=MS^2$ where S is the speed of strings. The atom's mass is being converted into mass fields.

When an atom absorbs a gravity propulsion field, its mass increases by the amount of energy it absorbs. ($E=MS^2$) When we have more energy going out than energy coming in, is when we lose matter. As gravity propulsion fields leave the universe, things slows down and matter is converted into mass fields. As things slow down, everything slows down including the rate at which we lose matter.

Energy Out = Energy In + Mass Converted To Energy

TGP predicts that in the past, time was faster, the speed of light was slower and mass was heavier. Here is a hypothetical example to illustrate this. In the past, if and when the speed of light was only 3,000 miles an hour (.84 miles per second), an atom would have 49,179,398,418 times more mass than it has now. That's about 50 billion times more mass. For each ounce of matter that we have remaining today, we have lost over 1,500,000 tons of matter. This has been converted into energy. Again, we don't notice it because everything is relative.

Because $E = MC^2$, total energy is the same now as it was then. It's just that mass and the speed of light have changed. The calculations below use the speed of light (C) because we don't know what the speed of S (Strings) is yet. S should be just a little faster than the speed of light. The difference in speed isn't important for this estimate.

Energy past	= Energy today
MC^2 past	= MC^2 today
M past x C^2 past	= M today x C^2 today
M past / M today	= C^2 today / C^2 past
	= (186,282 miles per second) ^{2}/ (.84 miles per second) ^{2}
	= 34,700,983,524 / .7056
	= 49,179,398,418
	= About 50 billion

"M past" is the mass of the past. "M today" is the mass today. The ratio "M past / M today" means each atom had 50 billion times more mass in the past than we have today.

7.7 Gravity Propulsion Changes Mass

According to TGP, as gravity propulsion force increases, mass increases. As gravity propulsion forces decrease mass decreases. This is similar to $E=MC^2$. The equation is:

$$\Delta P = \Delta MS^2$$

The change of gravity propulsion force (ΔP) equals change of mass (ΔM) times speed of strings (S) squared.

7.8 Ether Flow

This section explains what an ether flow is. Here is my definition:

"Ether flow is the resultant flow of gravity propulsion fields."

"Ether flow" is a short way of saying "resultant flow of gravity propulsion fields." For example, we can now say, "Where there is a mass, there is an ether flow towards it." We could say the same thing without referring to ether flows by saying, "Where there is a mass, the mass absorbs gravity propulsion fields, so there are more gravity propulsion fields flowing toward the mass than are flowing away from it, which produces a resultant flow of gravity propulsion fields towards it." If you do not like to use the term "ether flow" you can substitute the phase "resultant flow of gravity propulsion fields" instead. I'm trying to keep things simple.

When a mass travels with the ether flow, the ether flow accelerates the mass and it goes faster. When a mass travels against the ether flow, the ether flow decelerates the mass and it goes slower. The same thing applies to light. Light is composed of photons. Photons are small particles of mass. Ether flow pushes light just like it pushes any particle. We don't notice this effect much because photons are very small and travel very fast. When light travels with the ether flow, the ether flow accelerates the light and it goes faster. When a mass travels against the ether flow, the ether flow decelerates the light and it goes slower.

Here are other ways to describe how ether flow affects matter and light:
1. When light (or matter) travels by a mass, it will bend around the mass because the ether flow pushes it toward the mass.
2. Light (or matter) that travels directly towards a mass, it will speed up because it is going with the ether flow that accelerates it.
3. Light (or matter) that travels away from a mass, it will slow down because it is going against the ether flow that is decelerating it.

Here is the classical definition of ether, "Ether fills all space and is the medium that allows light to travel through space." This is true. I know that the word "ether" is an unpopular word at the time of this report. I hope it becomes popular. Ether is the best term to describe how forces affect and change time and space.

The table below compares the classical ether theory description to TGP's gravity propulsion fields.

Classical Ether Theory	TGP Gravity Propulsion Fields
Fills all space.	Fills all space.
When light travels against the ether flow, it slows down.	When light travels against resultant gravity propulsion fields, it slows down.
When light travels with the ether flow, it speeds up.	When light travels with resultant gravity propulsion fields, it speeds up.
All light travels through it.	All light travels through it.
Allows light to travel through space.	Allows light to travel through space.

Gravity propulsion fields are not ether. Gravity propulsion is just a part of ether. The Quantum String Theory gives a more accurate definition and explanation of ether in Section 9.1.

7.9 Speed Changes Throughout the Universe

Since TGP states that gravity propulsion fields are leaving the universe, there is a constant ether flow away from the center of the universe. This changes the speed of matter, the speed of light and the speed of time throughout the universe. The table below shows these differences.

	Center Of The Universe	Edge Of The Universe
1. Galaxies travel:	Slowest	Fastest
2. Speed of light:	Slowest inward	Fastest outward
3. Gravity propulsion:	Strongest	Weakest
4. Speed of time:	Fastest	Slowest

1. Near the edge of the universe, matter travels the fastest outward from the center of the universe. The ether flow from the center of the universe is constantly accelerating it. In the center of the universe, matter doesn't move. It is being pushed equally in all directions.

2. When light travels toward the center of the universe, it slows down. This is because light is traveling against the ether flow. This is like fish swimming upstream. This slowing is constant and can be significant.

    As light travels outward from the center of the universe it speeds up a little. This is because light is traveling with the ether flow. In terms of superstrings, photons (light) are a small group of strings. A group of strings does not travel as fast as a single string. Therefore, photons travel close to but not as fast as single strings. Photons have potential to increase their speed to that of single strings as they travel outward from the center of the universe. However, this increase of speed is very slight. This is because single strings do not travel much faster than light (photons).

3. In the center of the universe, almost all gravity propulsion fields are bouncing back and forth. Very little of the fields leave the universe never to return. This provides strong gravity propulsion forces from all directions on matter. At the edge of the universe, there are very weak gravity propulsion forces pushing matter outward and no gravity propulsion forces pushing inward.

4. The stronger gravity propulsion fields are, the faster atom's electrons will orbit. The strong gravity propulsion forces at the center of the universe, makes the speed of time go the fastest. The weaker gravity propulsion fields are, the slower atom's electrons will orbit. The weak gravity propulsion forces at the edge of the universe, makes the speed of time go the slowest.

7.10 Light Shifts

Light changes frequency, speed and momentum based upon differences between the light source and the destination. TGP predicts that light changes as it travels due to the following three basic factors:
1. Doppler Shifts
2. Time Shifts
3. Speed Shifts

Larry James

7.10.1 Doppler Shift

The Doppler Shift is the apparent variation in frequency as the source of the frequency approaches or moves away from an observer. As the distance between the source and observer increases, the pitch or frequency becomes lower. As recession velocity increases, light frequencies become redder. As the distance between the source and observer reduces, the pitch or frequency becomes higher. As approaching velocity increases, light frequencies become bluer. This effect was named after Christian Johann Doppler, an Austrian physicist, who first wrote about this principle in 1842. The Doppler shift applies to source movement of sound frequencies, light frequencies and radio frequencies.

7.10.2 Time Shift

TGP makes the following prediction and definition:

The Time Shift is the apparent variation in frequency due to the difference in the speed of time between the source and the destination.

When light comes from a slower to a faster speed of time, light frequency is shifted lower (redder). When light comes from a faster to a slower speed of time, light frequency is shifted higher (bluer).

Here are some examples. Gravity slows down the speed of time. A star's emitted light is shifted lower and redder because its gravity slows down its speed of time. On earth, the incoming star's light is shifted higher and bluer because the earth's gravity slows down the speed of time of the viewer. Because the star's gravity is greater than the earth's gravity, the net difference in the speed of time shifts the star's light frequency to be redder. The change of frequency due to the time shift is $\Delta G/P_o$.

Time goes faster in the center of the universe and goes slower the further it is from the center. A star in the center of the universe has a faster speed of time than on earth. Because of this time difference, the star's emitted light frequency will be shifted higher and bluer when viewed from earth. A star near the edge of the universe has a slower speed of time than on earth. Because of this time difference, the star's emitted light frequency will be shifted lower and redder when viewed from earth.

7.10.3 Speed Shift

TGP makes the following prediction and definition:

The Speed Shift is the apparent variation in effective frequency due to changes in the speed of light.

When light slows down, it reduces its momentum and kinetic energy and thereby reduces its effective frequency. When light speeds up, it gains momentum and kinetic energy and thereby increases its effective frequency.

TGP predicts that ether flow affects the speed of light. When light travels with the ether flow, it increases the light's speed, momentum, kinetic energy and effective frequency. When light travels against the ether flow, it decreases the light's speed, momentum, kinetic energy and effective frequency. A star's light coming from the center of the universe to earth will increase in speed because it traveled with the ether flow. A star's light coming from the outer part of the universe to earth will decrease in speed because it traveled against the ether flow.

TGP predicts that gravity affects the speed of light. Light is composed of particles (photons). Like all particles, gravity affects photon particles the same way. When light travels away from a star, the star's gravity slows it down, which reduces the light's speed, momentum, kinetic energy and effective frequency. When light travels towards the earth, the earth's gravity increases the light's speed, momentum, kinetic energy and effective frequency.

The following explains equations for changes to the speed and frequency of light. For the following explanation, imagine that a beam of light is shooting straight up and it is measured at a height (h). Gravity slows it down as it travels up path h. First, lets explain the variables.

h = Height light travels.
v_0 = Velocity of light at the source.
Δv = Change of speed of light over path h.
c = Average speed of light over path h. $c = v_0 + \Delta v/2$
g = Average gravity over path h.
Δf = Change of frequency over path h.
Δg = Change of gravity over path h.
Δt = Change of time over path h.
P_0 = Gravity Propulsion force.

Now that we have defined the above variables, we can derive the equations.

h/c = Time it takes light to travel over path h.

Multiply this times the average gravity (g) to get the change of speed (Δv):

$$\Delta v = -gh/c$$

The negative sign means that downward gravity reduces upward velocity and increases downward velocity. Divide this by its initial velocity (v_o) to get the change of frequency (Δf):

$$\Delta f = \Delta v/v_o = -gh/cv_o$$

To be more mathematically accurate, we should integrate g and c over h because g and c do not change linearly. The above equations define the change of speed and frequency of light relative to the source. To be relative to the destination at h, we need to factor back in the change to the speed of time there, which is:

$$\Delta t = -\Delta g/P_o$$

Relativity: Einstein's Relativity Theories do not recognize that gravity changes the speed of light. For example, Relativity requires that we use the speed of light in a vacuum for c and v_o in the above equations. In addition, light changes speed based on its frequency through a medium (like air) as well as by gravity. These are areas are where Relativity becomes inaccurate. TGP separately explains physical effects and time effects and thereby can account for all effects. Relativity cannot do this.

The Pound-Snider experiment[8] measured changes to the frequency of light as it traveled up and down 75 feet. Unfortunately, they only reported the sum of these two frequency changes. They did not separately report the frequency change going up and the frequency change going down that they measured. TGP predicts that the frequency change going up would be slightly more than the frequency change when going down because gravity had more time to slow the light down as it traveled upward the same distance. The report did make the following conclusion that is in agreement with TGP predictions, "It is to be noted that no strictly relativistic concepts are involved and the description of the effect as an 'apparent weight' of photons is suggestive. The velocity difference predicted **is identical** to that which a material object would acquire in free fall for a time equal to the time of flight." In other words, photons (light) are particles that gravity affects like all other objects (including apples).

---

[8] R.V. Pound and J.L. Snider, "Effect of Gravity on Gamma Radiation", Phys. Review, vol 140, issue 3B, 788-803 (1965).

7.11 Cosmic Red Shift

This section gives examples of frequency and momentum shifts that affect light from distant galaxies. A galaxy emits or absorbs various wavelengths of light more strongly than others. Based on this, astronomers have observed that galaxies produce a various amounts of red shift. The "red shift" provides evidence that our universe is expanding. Based on this red shift, astronomers tried to figure out the distance to galaxies.

In 1929, Edwin Hubble reported that the universe is expanding at a **constant** rate. Hubble's law says that recession velocity is directly proportional to the distance of the object. Specifically, the recession velocity (v) of a galaxy is equal to its distance (D) multiplied by a quantity called Hubble's constant (H_o). The "o" is pronounced "naught". H_o means the current value, since the Hubble "constant" changes with time. H_o is the ratio of velocity to distance as the universe expands.

Based on Hubble's law, astronomers could easily calculate how far away it was. They simply multiply the red shift by Hubble's constant. The problem is we now know that the universe expansion is **accelerating** (Section 7.2) and that the Time Shift and Speed Shift affect the observed red shift (Section 7.10). This means that the Hubble constant is not a simple constant.

Figure 23 below shows the various light shifts that occur depending on whether the galaxy is towards the center of the universe or towards the edge of the universe. Try to visualize how the photon's speed and speed of time changes relative to its surroundings.

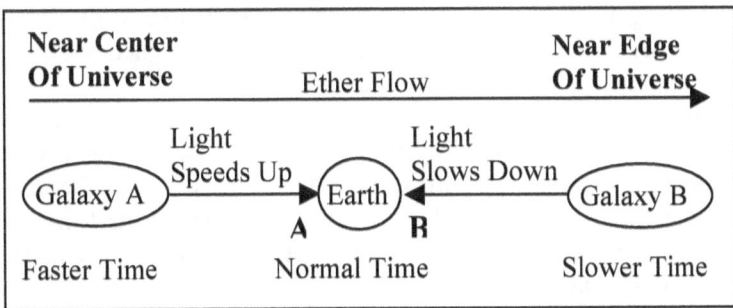

FIG. 23. Cosmic frequency shifts relative to Earth.

Point A. The following are changes to Galaxy A's light frequency as it travels to Earth:

1. Doppler Shift - Red. Frequency shifts lower because the Earth is moving away from galaxy A. This is also called "Recession Red Shift".
2. Gravity Time Shift - Red. The star's light frequency shifts lower because the star's gravity reduces time.
3. Gravity Speed Shift - Red. As light leaves the star, the star's gravity, the star's solar system gravity and the star's galaxy gravity slow the light down. This shifts the light's effective frequency lower.
4. Ether Speed Shift - Blue. Effective frequency shifts higher because light speeds up as it travels with the ether flow. Because the speed of light starts out near its maximum potential speed, the increase of light speed is minimal.
5. Cosmic Time Shift - Blue. Frequency shifts higher because the source is in a faster time, towards the center of the universe, and the destination (earth) is in a slower (normal) time.
6. Gravity Speed Shift - Blue. As the light reaches Earth, our galaxy's gravity, our solar system's gravity, the Sun's gravity and Earth's gravity, increases the speed of incoming starlight. This increases the light's effective frequency.
7. Gravity Time Shift - Blue. Incoming light frequency shifts higher because Earth's gravity reduces the speed of time.

Point B. The following are changes to Galaxy B's light frequency as it travels to Earth:

1. Doppler Shift - Red. Frequency shifts lower because Galaxy B is moving away from earth. This is also called "Recession Red Shift".
2. Gravity Time Shift - Red. The star's light frequency shifts lower because the star's gravity reduces time.
3. Gravity Speed Shift - Red. As light leaves the star, the star's gravity, the star's solar system gravity and the star's galaxy gravity slow the light down. This reduces the light's effective frequency.
4. Ether Speed Shift - Red. Effective frequency shifts lower because light slows down as it travels against the ether flow. This shift can be significant.
5. Cosmic Time Shift - Red. Frequency shifts lower because the source is in a slower time, towards the edge of the universe, and the destination (earth) is in a faster (normal) time.
6. Gravity Speed Shift - Blue. As the light reaches Earth, our galaxy's gravity, our solar system's gravity, the Sun's gravity and Earth's

gravity, increases the speed of incoming starlight. This increases the light's effective frequency.

7. Gravity Time Shift - Blue. Incoming light frequency shifts higher because Earth's gravity reduces the speed of time.

7.12 The Center of the Universe (Predictions 9 to 12)

TGP requires that there is a center of the universe. TGP predicts that eight unique verifiable cosmic events exist and that these events are aligned to each other.

	Towards Center of the Universe	Opposite Direction
1. Galaxy Density	Highest	Lowest
2. Supercluster Distance	Closest	None
3. Speed of Time	Fastest	Slowest
4. Background Radiation	Highest	Lowest

7.12.1 Galaxy Density (Prediction 9)

TGP predicts that the direction towards the center of the universe has the highest average density of galaxies of any area in the universe. This is because galaxies near the center of the universe are not moving away as fast as other galaxies. This area should appear as an outward line of galaxy superclusters. In the opposite direction, the density of galaxies will be lower. The average density of galaxies should be lowest in that direction.

If the universe did not begin with a big bang, then galaxies and would be evenly distributed throughout the universe and there would be no galaxy superclusters.

Confirmed Verification: John Huchra and Margaret Geller lead a ten-year project from 1985 and 1995 called "CfA Redshift Survey"[9]. This project measured the red shifts of about 18,000 bright galaxies in the northern sky. Out of these 18,000 galaxies, they found an area that has a highest density cluster of galaxies. This is called the "Virgo cluster of galaxies" because it is in the constellation Virgo. The center of the universe is in this general direction.

---

[9] See http://cfa-www.harvard.edu/~huchra/zcat

7.12.2 Local Supercluster (Prediction 10)

TGP predicts that the direction towards the center of the universe should have a supercluster of galaxies that appears significantly closer to us than any other supercluster. This supercluster appears close to us because their speed of time is faster. See point A in the previous section. In the opposite direction, there will not be a supercluster. If the universe did not begin with a big bang, then galaxies and any galaxy superclusters would be evenly distributed in distance from us.

Again, we see research that supports this prediction. The "CfA Redshift Survey" found a supercluster of galaxies that appears closest to us. This is called the **"Local Supercluster"**. The Local Supercluster is near the center and the densest area of the Virgo cluster of galaxies. It is in the head of the constellation Virgo. There is no similar Local Supercluster anywhere else. The Local Supercluster is moving "towards" us at the rate of 170 Km/s (relative galactic speed) while the earth is moving "away" at the rate of 360 Km/s (relative galactic speed). Movement is relative as explained in the section "Background Radiation."

The Local Supercluster is in the same direction as the Virgo cluster of galaxies. This is the general direction to the center of the universe.

7.12.3 Cosmic Speed of Time (Prediction 11)

TGP predicts that the direction towards the center of the universe has the fastest speed of time. This is where the gravity propulsion fields are the strongest. In the exact opposite direction, the speed of time is the slowest. This is where gravity propulsion forces are the weakest. If TGP were not true, then the speed of time would be the same throughout the universe.

This prediction has been verified by measurements of radiance curves of Type Ia supernovae. The theory of Gravity Propulsion is the only theory that explains the differences in the speed of time found in Type Ia supernovae radiance curves.

All Type Ia supernovae have the two things in common:
1. They always radiate the same amount of peak light.
2. Their increase and decrease in radiance over time are the same.

All Type Ia supernovae radiate the same brightness because they explode when they obtain the same amount of critical mass. They begin as a white dwarf star. A white dwarf star is about the size of the Earth but has about the same mass as our sun. This dwarf star has another star orbiting it. The dwarf star accumulates matter from the companion star until its intense gravity triggers a runaway thermonuclear explosion. Since all Type Ia supernovae have the same amount of mass when they explode, they give off the same amount of light over time.

All Type Ia supernovae increase and decrease in radiance at the same rate of time. The radiance of the explosion grows brighter over about 20 days, and then it fades over the following months. Their brightness over time is called a radiance curve or light curve. All Type Ia supernovae have the same radiance curve.

TGP predicts that near the center of the universe, time goes faster. Therefore, in that direction, we should see Type Ia supernovae that are further away, reach peak brightness faster and dim faster than nearby ones. Their radiance curve should be more compressed. Their radiance curve's compression factor is directly proportional to their increase in the speed of time.

In the opposite direction, TGP predicts that near the edge of the universe, time goes slower. Around there, we should see Type Ia supernovae reach peak brightness later and dim later. Their radiance curve is expanded. Their radiance curve's expansion factor is directly proportional to their decrease in the speed of time.

The Supernova Cosmology Project has found over 75 Type Ia supernovae. They have measured their radiance curves. They discovered that their radiance curves are not the same. The following graph[10] shows these differences.

The vertical axis measures brightness magnitude graphed as a log scale $M_v - 5\log(h/65)$. The horizontal axis measures number of days before and after the peak brightness.

---

[10] Poster displayed at the American Astronomical Society meeting in Washington, D.C., January 9, 1998 (Perlmutter et al., B.A.A.S., v. 29, no. 5, p. 1351, 1997).

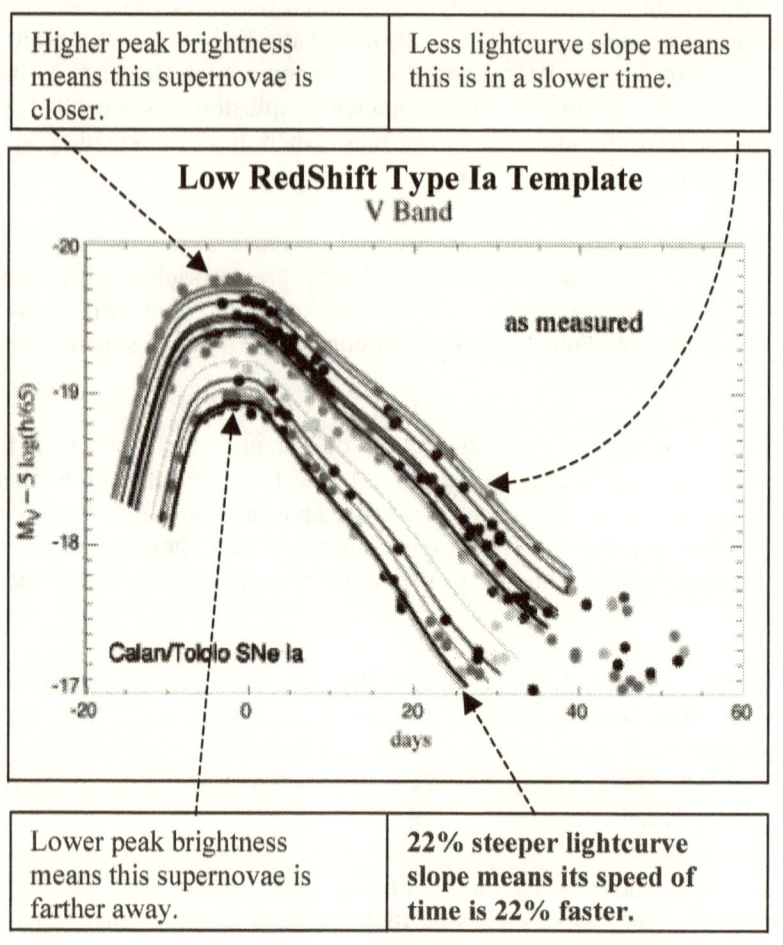

| Higher peak brightness means this supernovae is closer. | Less lightcurve slope means this is in a slower time. |

Low RedShift Type Ia Template
V Band

as measured

Calan/Tololo SNe Ia

| Lower peak brightness means this supernovae is farther away. | 22% steeper lightcurve slope means its speed of time is 22% faster. |

FIG. 24. Graph of Type Ia brightness curves showing that the speed of time is faster towards the center of the universe.

The above graph was from www-supernova.lbl.gov/public/papers/ aasposter198dir/wwwposterId.jpg on 5/27/2002. This graph was created by C Pennypacker, M. DellaValle, R. Ellis, R. McMahon, B. Schaefer, P. Ruiz-Lapuente and H. Newberg. I added the arrows and comments. Below this graph were these words, "Type Ia supernovae observed 'nearby' show a relationship between their peak absolute luminosity and the timescale of their light curve: **The brighter supernovae are slower and the fainter supernovae are faster.** (see Phillip, Ap.J.Lett., 1994 and Riess, Press; & Kirshner, Ap.J.Lett, 1995)"

These supernovae are in the general direction toward the center of the universe. This is where galaxy clusters are the densest. In order to find Type Ia supernovae, astronomers take pictures of galaxy clusters. About two weeks later, they take the same pictures again and compare the pictures to see if brightness at any point has increased. Any increase in brightness is the beginning of a supernovae explosion. Type Ia supernovae last only a few days and are extremely rare. Astronomers have to scan hundreds of galaxies in order to find one Type Ia supernovae. This is why they look at areas that have the most galaxies. This is in the direction of the Virgo cluster of galaxies. This is also the general direction towards the center of the universe.

In conclusion, astronomical data has confirmed that time goes faster towards the center of the universe just as TGP predicts and no other theory explains.

7.12.4 Background Radiation (Prediction 12)

TGP predicts that the direction towards the center of the universe must have the highest background radiation level. This is where the gravity propulsion fields are the strongest. The exact opposite direction must have the lowest background radiation. This is where the gravity propulsion fields are the weakest. If the universe did not begin with a big bang, then any background radiation would be evenly distributed throughout the universe.

Verification Status: Between 1989 and 1993, a National Aeronautics and Space Administration's (NASA) satellite named the Cosmic Background Explorer (COBE) mapped the cosmic background radiation (CBR). The average background temperature is about 2.725^0 Kelvin (degrees above absolute zero). They also found that the temperature varies by about $.003353^0$ K between two opposite points in the sky. The average temperature gradually changes between these two points. They call this the dipole anisotropy. "Dipole" means there are two poles. "Anisotropy" means opposite polarization, charge or variation.

Figuring the direction of the universe's center based on the dipole anisotropy is complicated. This difference in cosmic temperature can be interpreted two ways depending upon the assumptions. There are the two assumptions:
1. The average cosmic temperature is the same everywhere. This is conventional thinking.

2. The average cosmic temperature is different. This is the view of TGP.

In the past, scientists have assumed that the average cosmic temperature is the same everywhere. Therefore, as the earth moves through it, the temperature appears different only because of the Doppler effect. As the earth moves away from a point, that point appears cooler. And, as the earth moves towards a point, that point appears hotter. They then calculate the apparent speed of the earth towards and away from these points.

Assuming that cosmic background radiation is the same everywhere, the earth is moving "towards" the hottest point located in the direction $(l=264.26^0, b=48.22^0)$ $(a=11^h\ 12.2^m, d=-7.06^\circ)$. The earth is moving towards this point at about 360 Km/s (galactic speed). This assumes that only our speed causes the temperature difference of $.003353^0$ K.

However, TGP predicts that the average cosmic temperature is different. Cosmic temperature is hotter near the center of the universe and is cooler in the opposite direction. **This means that the earth may be moving in the opposite direction than is currently assumed.** The earth may be moving away from the hotter point instead of towards it. As the earth moves away from this hotter point, the point appears less hot than it actually is. As the earth moves towards the cooler point, that point appears hotter than it actually is. Thereby, the relative speed of the earth is different than is currently calculated.

The direction that the earth is moving is not exactly along a line that points to the universe center. This is because bodies revolve. The earth revolves around the sun at about 30 km/s. The sun revolves around the center of our Milky Way galaxy. Our galaxy revolves around our local supercluster. All of these movements have to be understood before we can deduce the exact direction to the center of the universe based only on our dipole anisotropy. The velocity of our local group of galaxies relative to the Cosmic Background Radiation Rest is about 600 km/s in the direction of $(a=10.5^h, d=-26^\circ)$ $(l=268^\circ, b=27^\circ)$.[11]

---

[11] www.circlon.com/HTML/gravpotenergy.html, 4/10/2002

7.13 The Cosmological Constant Isn't

In 1917, Einstein introduced the term "cosmological constant" and used the Greek letter "Lambda" to represent it. The value of Lambda represents the rate that empty space is created or the rate it is destroyed. The value that Einstein gave Lambda balanced his equations of General Relativity. Lambda's initial value presented a stable universe that would neither expand nor collapse on itself.

Einstein's cosmological constant was a good idea at that time. We now know there is something that is constantly pushing everything apart. However, it isn't the creation of empty space that is doing it. And, it isn't the same everywhere. Its not like blowing up a balloon that has dots on it and seeing the dots separate. Einstein's cosmological constant requires that there is no center of the universe. Current evidence shows that there is a center of the universe that TGP predicts. The cosmological constant is not a constant. It's an equation that has yet to be derived. This is explained next.

7.14 The New Cosmological Equation

Cosmology is the theory of the universe as a whole and of the general laws that govern it. This includes the universe's distant past and its future. The cosmological equation describes the following:
- How fast time, matter and the force of gravity propulsion change over time.
- How time, matter and the force of gravity propulsion vary throughout the universe from a relative position and time.

The cosmological equation provides the foundation to figure out the following:
- Where the center of the universe is.
- The true distances and speeds of galaxies around us.
- How fast the universe is expanding and accelerating.
- How old the universe is.
- How far we are from the center of the universe.

The cosmological equation provides a macro (high-level) model of the universe. A major assumption is that all matter is equally distributed around a point. It does not account for smaller effects like vortexes where galaxies, including our own, revolve around each other. It does not account for dense clusters of galaxies and matter that would reflect back more than average amounts of gravity propulsion fields. It does not account for areas that are less dense than normal that produce less than average gravity propulsion fields. Still, the cosmological equation provides the relative framework for understanding how time and space change since time began and how it varies throughout the universe.

The cosmological equation problem statement is given in Appendix A.

7.15 How Old Is the Universe?

What is our time (T) in the cosmological equation? We can derive this from the slope and intensities of Type Ia supernovae radiance curves. Every point on the cosmological exponential time curve can be uniquely identified by the change of the change of slope. Thus, we can identify when we are by the change of the change of the speed of time. By finding the change of the change of the speed of time, we can determine when we are on our cosmological time curve and how old the universe is.

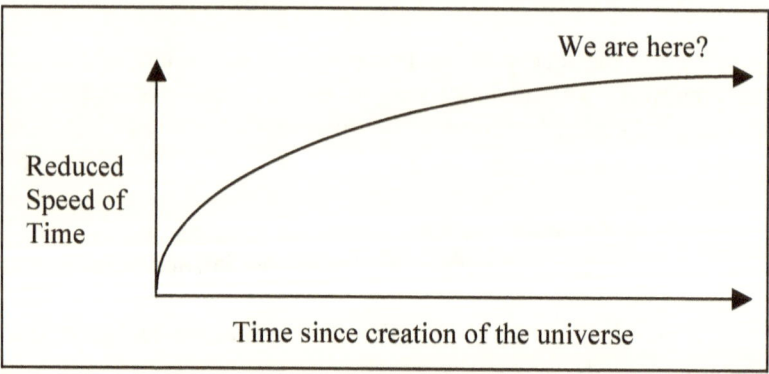

FIG. 25. Example of exponential time curve.

TGP predicts that Type Ia supernovae's distance, from us to the center of the universe, is not perfectly linear to its radiance curve's compression factor (change of time). The slight difference in linearity should be an exponential curve. For example, the furthest Type Ia supernova from us, as shown previously, is going 22% faster than we are. Lets says there is a Type Ia supernovae that is exactly half way in between us that is going 10.9% faster than us. If the speed of time was dependent only on distance then it should be going 11% faster. The .1% difference is because of the exponential curve. By accurately identifying these differences, we may be able to determine where we are on the exponential curve. By identifying this exponential curve and matching it to the cosmological equation's curve, we should be able to determine how old the universe is.

7.16 The Universe Will End

TGP predicts that the universe will end at some point in time. The universe will never reverse its expansion. It will never collapse back. This is a one way trip that will end. This is because there is no force that attracts matter together. Mass does not attract mass. The universe is not "closed". This means that no amount of critical mass will ever cause the universe to reverse expansion and collapse into a big crunch. Nothing will slow down the expansion of the universe.

Time slows down asymptotically as gravity propulsion forces slowly get closer to zero. Someday, it may take a billion years of our current time, to advance one second in the far future. But time goes on.

Currently space gas collects into larger masses. This process can build new stars into a critical mass so they can have fusion and glow. But the universe expansion is accelerating, due to gravity propulsion. The disbursal of matter will continue to increase faster than accretion (collection) of matter. Someday, stars will run out of the critical mass they need to maintain fusion. All stars will burn out. At that point, the universe is dead, as we know it.

7.17 Questions

The following questions are about TGP as stated in this section. The answers are in Section 16.

7.1   Why is the expansion of the universe accelerating?
7.2   Where are gravity propulsion forces the strongest?
7.3   In the center of the universe, is the speed of time the same as ours, faster or slower?
7.4   In the center of the universe, are galaxies about the same average distance, closer together or further apart than galaxies near us?
7.5   On earth, is the gravity propulsion force per cubic inch staying the same, getting weaker or getting stronger?

7.6 On earth, is the mass of an average particle staying the same, getting less or growing?

7.7 On earth, is our speed of time staying the same, slowing down or speeding up?

7.8 On earth, what is happening to our speed of light?

7.9 Name the two natural forces that increase the speed of starlight.

7.10 Why do some Type Ia supernovae reach peak brightness faster then dim faster than other Type Ia supernovae?

8.0 TGP Compared to Other Theories

In this section, we briefly compare the Theory of Gravity Propulsion to other theories. The only thing that TGP conflicts with is other theories. TGP does not conflict with any facts. TGP differs from all other competing theories in various ways. This section identifies some of the differences to help you compare them. These theories are listed in order that they were developed.

TGP is based on a more elemental physics than all other theories. It is not an extension to any current or previous theory. In concept, it comes closest to the superstring theory. TGP is a new physics. TGP more than "upsets the physic's apple cart." It blows it apart into quantum pieces and builds it back together again. The result is that TGP explains the basic principles that other theories were based on. It reveals the "magic" behind them.

8.1 Michelson-Morley Experiment

In 1887, Albert Michelson and Edward Morley performed an experiment that supported the theory that ether did not exist and that light travels at a constant speed. They created an experiment to measure ether flow. If ether existed, then light must slow down when it travels against the ether flow. Also light must speed up when it travels with the ether flow. Their experiment was designed to measure the motion of the earth through ether. They created an interferometer to measure small differences in wavelengths as light traveled at right angles to each other. The experiment's conclusion was that light traveled the same speed no matter which direction they rotated the beams of light. They concluded (falsely, I will argue) that ether flow did not exist and therefore ether did not exist. Because of their experiment, for over a hundred years, the scientific community assumed ether did not exist and that light travels at a constant speed.

TGP differs from the Michelson-Morley conclusions in the following ways:

Michelson-Morley Experiment	Theory of Gravity Propulsion
The speed of light is a constant.	The speed of light is not a constant.
Ether does not exist.	Ether exists.

The problem with the Michelson-Morley experiment is that they only measured the speed of light horizontally. If they compared the speed of light vertically they would have discovered that light travels faster downward and slower upward (Section 7.10.3).

8.2 Planck's Theory

In 1900, the German physicist Max Planck concluded that the energy of a quantum of light is equal to the frequency of the light (v) multiplied by Planck's constant (h). The value of Planck's constant is h = 6.626×10^{-34} joule-second. Planck's theory states that energy can be emitted or absorbed by matter **only** in small, discrete units of "quanta". We now call these quantas of light "photons". We now know that photons are particles. Photons are particles emitted by electrons. An electron absorbs a photon when the photon impacts the electron.

TGP differs from Planck's theory. TGP requires photons to be composed of even smaller quantum strings. TGP also discovered that, **"Individual quantum strings transmit pure energy."** Electrostatic fields, magnetic fields and gravity fields are pure energy. These fields are properties of quantum strings. These forms of energy are not transmitted by photons or by any discrete units of "quanta" as defined by Planck's theory. Individual quantum strings are not discrete units of energy. They are pure energy. This is explained in section 9.0 "TGP's Quantum String Theory."

8.3 Theory of Special Relativity

In 1905, Albert Einstein introduced his Theory of Special Relativity. He developed this theory to explain the results of the Michelson and Morley experiment in 1887. He also based his theory on the Planck's Theory of 1900 that stated that all energy is transmitted by photons (or quantas).

The Theory of Gravity Propulsion explains how The Theory of Special Relativity works in terms of quantum strings. There are also differences. The following table shows the main differences between these theories.

Theory of Special Relativity	Theory of Gravity Propulsion
The speed of light is a constant.	Speed of light changes due to medium, gravity and ether flow.
Nothing goes faster than the speed of light (C).	The speed of individual quantum strings (S) is faster than light.
Does not explain how mass and energy are converted.	Explains how mass and energy are converted.
Time slows down as it approaches the speed of light.	Time slows down as it approaches the speed of strings.
Does not explain what causes time.	Explains what causes time.
Mass increases to infinity as it approaches the speed of light.	Mass increases to infinity as it approaches the speed of strings.

The Theory of Special Relativity was wrong because it was based on the wrong conclusions of the Michelson-Morley experiment and Planck's Theory.

Some people still have trouble accepting that the speed of light is NOT a constant. Light slows down when it travels through a denser medium. When light travels from one medium into another it bends because the speed of light changes. A material's index of refraction (n) is the ratio of the speed of light in a vacuum ($c = 3 \times 10^8$ m/s) to the speed of light in that material (v). $n = c/v$.

TGP predicts that quantum strings go faster than light. This is explained in more detail in the section "Quantum Strings Travel Faster Than Light." Here is an overview. Light is composed of photons. Photons are particles. Photons require a force to accelerate them to the speed that they go. This force has to go faster than photons do because photons can never go as fast as the force that pushes them. Individual traveling quantum strings are that force. These strings travel slightly faster than the speed of light (photons).

8.4 Theory of General Relativity

In 1916, Albert Einstein introduced The Theory of General Relativity to explain gravity. The main concept of the Theory of General Relativity is that gravity is like an elevator. A person feels the same downward force on Earth as if they were in an elevator in space accelerating smoothly at about 32.2 feet per second squared. Also a person floating in space feels the same weightlessness as if they were in an elevator on Earth dropping in free fall (except for the sudden stop). This is AS IF gravity changes time and space. Einstein declared that gravity DOES change time and space. TGP challenges this conclusion. This is like saying, "Gravity is like an elevator and therefore gravity is an elevator."

The Theory of Gravity Propulsion explains how the Theory of General Relativity works in terms of quantum strings and then goes beyond the limitations imposed by Einstein's theories. The table below shows some of the differences between these theories.

Theory of General Relativity	Theory of Gravity Propulsion
Gravity field: Mass creates a gravitational field that curves four-dimensional space-time. The gravitational field has energy but it does not transport energy.	Gravity field: Mass radiates a mass field. Mass fields are composed of traveling strings.
Does not explain how mass changes space-time to create gravity.	Explains how gravity works.
Does not explain how other attractive forces work.	Explains how all attractive forces work.
Space is expanding, contracting or staying the same. Space cannot accelerate expansion or contraction.	Universe expansion is accelerating. There are no other options.
There is no center of the universe.	There is a center of the universe.
Galaxies must be evenly distributed throughout the universe.	Galaxies are denser in the center of the universe. There are large voids in the universe.
When gravity bends light, its speed stays the same.	When gravity bends light, its speed increases.
The speed of time changes due to speed and gravity.	The speed of time changes due to speed, gravity and gravity propulsion.

8.5 Quantum Mechanics Theory

In 1924, French physicist Louis Victor, Prince de Broglie, discovered that matter has both wave and particle properties. The wavelength of matter waves is given by the equation $\lambda = h/mv$, where h is Planck's constant, m is the particle mass and v its velocity. Electron orbits can then be described as differential wave equations. Soon after, various scientists developed Broglie's idea into various wave equations to describe how electrons orbit the atom.

- In 1926, Erwin Schrödinger developed the Schrödinger wave equation that described the wave properties of electrons in atoms.
- In 1928, British mathematical physicist Paul Adrien Maurice Dirac added relativity to the wave equations of electrons. He declared that electrons also have spin and magnetic properties.

TGP and Quantum Mechanics Theory differ by how they describe electrons. The table below shows some of these differences.

Quantum Mechanics	The Theory of Gravity Propulsion
Some electrons orbits (e.g.: p, d and f) can travel through the nucleus.	Electrons orbit around the nucleus. Electrons never go through the nucleus.
Electrons have spin and magnetic poles.	Explains why electrons can have magnetic poles.
Electrons behave as waves.	Electrons are particles that radiate mass fields. Mass fields have wave properties.
Does not explain why electrons are attracted to the nucleus.	Explains why electrons are attracted to the nucleus.

With few exceptions, Quantum Mechanics wave equations have accurately predicted electron orbital energies in atoms when compared to experimentation. In comparison, TGP states that electrons are not pure waves. TGP requires electrons to be physical particles that physically orbit atoms in the classical sense. In 1913, Danish physicist Niels Bohr declared that electrons orbit the nucleus in quantum levels. In 2000, Carl Segor showed that by adding relativity to the Bohr model, we can accurately predict electron orbital energies in the hydrogen atom.[12]   Therefore, Quantum Mechanics' is not the only way to calculate electron orbitals. Electrons do orbit the nucleus, as TGP requires.

Quantum Mechanic provides a high level practical approach to modeling the atom. TGP is a more elemental theory requiring more complex math to model the atom. Although TGP could be used to model atoms, TGP can be better used to model and understand fields and forces.

8.6 Casimir Effect Theory

Hendrik Brugt Gerhard Casimir, while at Philips Research Laboratories, studied attractive forces between two conducting plates in a vacuum. This study was to measure Van der Waals attractive forces among molecules. He published his findings to the Royal Dutch Academy of Science in 1948.[13] The equation of this attraction was:

$$F = \Pi^2 hcA/240d^4$$

Where:
  $\hbar$ = Planck's constant/2Π
  Π = Pie
  c = Speed of light
  A = Area of plate
  d = Distance between plates

There have been various theories to explain this attraction. Here is one of the popular theories. "The Casimir effect occurs because of omnipresent electromagnetic waves that exist in the vacuum of space. In between the plates, electromagnetic waves with wavelengths larger than the plate separation cannot exist. This imbalance of electromagnetic forces serves to push the plates together."

The Theory of Gravity Propulsion disagrees with the above theory. The Casimir effect does not require "omnipresent electromagnetic waves in the vacuum of space". Here is what causes the Casimir effect:

---

[12] "A Description for Discrete Electron Orbits about the Atomic Nucleus", by Carl Segor, June 19, 2000. See http://www.orbits.00space.cocm/electronorbit.html.
[13] H.B.G. Casimir, Proc. Ken. Ned. Akad Wetensch B51,793 (1948)

The primary cause of the Casimir effect is alternating resonating molecular dipoles.

Temporary dipoles exist between molecules. This is called the "London Force" and the "Dispersion Effect". Dipoles happen because electrons fluently move around in their orbits. Electrons are not evenly distributed in their orbits. Electrons seek positions and orbits of least resistance to the electrostatic forces around them. In a solid mass, molecular dipoles align up and synchronize their orbital orientations to paths of least resistance. Depending on the polarizability of the molecule, attraction can vary as much as $1/r^6$.

Figure 26 below illustrates how molecule dipoles align in a mass at one moment. The next moment, their dipoles will be reversed. This causes molecules to attract each other. Each oblong circle represents a molecule.

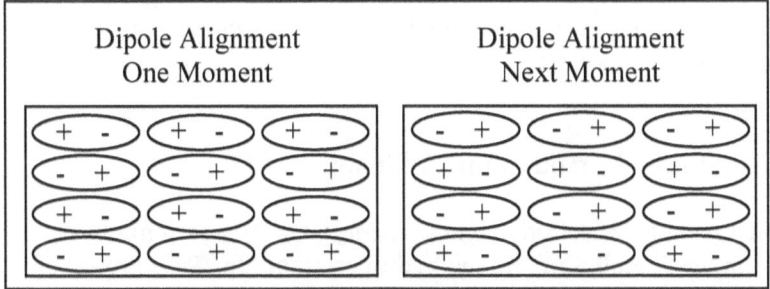

FIG 26. Molecule dipole movement in a solid.

Each mass, including plates, has their own alternating resonating molecular dipole movements. When two plates are brought close together, they become more like a single mass. As they become more like a single mass, their separate alternating molecular dipole movements become more synchronized to each other. The amount that they are synchronized to each other determines the amount of their mutual attraction.

Figure 27 below shows attraction between two plates due to dipole orientation at a given moment.

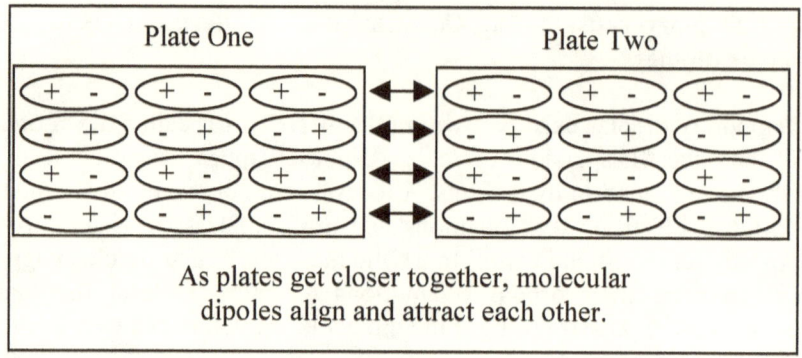

FIG. 27. Synchronization of molecular dipole moments of two close solid objects creates attraction.

The amount of attraction between two non-charged plates in a vacuum depends on their distance and their materials. Resulting Van de Waals attractive forces can vary between $1/r^3$ to $1/r^6$ depending on the material's static and dynamic polarities.

8.7 Quantum Electrodynamics Theory (QED)

The Quantum Electrodynamics Theory (QED) was initially developed to explain the "Lamb Shift" in the hydrogen atom. In 1947, Professor Willis E. Lamb and Professor Polykarp Kusch discovered the Lamb Shift. They discovered that some hydrogen electron orbital states were slightly greater than that predicted by the Dirac theory.[14] The difference is about 4.372×10^{-6} eV or 1057.852 Mhz. In 1955, they both received the Nobel Prize in Physics for this discovery. Since then others have verified this and found that the Lamb Shift effect exists in other atoms too.

QED Theory has concluded that other particles are pushing electrons outwards as they orbit the atom. QED Theory states that these additional particles pop in and out between the atom's nucleus and the electron's orbit. Some of these additional particles are photons and electron-positron pairs. Photons push electrons to create orbital oscillations. Electrons are also constantly pushed by high-energy (virtual) electron-positron pairs that

---

[14] W. E. Lamb, Jr., and R. C. Retherford, "Fine Structure of the Hydrogen Atom by a Microwave Method," Phys. Rev. 72, 241 (1947).

spontaneously fluctuate in an out of the vacuum of space at the rate of a billion trillion (1×10^{-21}) times a second.[15]

Note: Positrons are anti-matter particles. When a positron and electron meet, they usually annihilate each other and turn into pure energy.

Theory of Gravity Propulsion agrees with QED in that other particles are pushing electrons outwards as they orbit the atom. And, that these particles come in between the nucleus and the electron orbit. Where TGP differs with QED is that TGP states that these particles are invisible. These particles are not matter and anti-matter pairs vibrating in and out of existence. Interactions between protons, electrons and invisible particles should exactly match the various measured Lamb Shift effects. This is explained in section 10.6 "The Lamb Shift Effect".

8.8 Quantum Chromodynamics Theory (QCD)

In 1963, American physicists Murray Gell-Mann and George Zweig proposed the concept of quarks to explain a class of elementary particles called hadrons. Protons and neutrons are included in this hadrons classification. This became known as the Quantum Chromodynamics Theory, or QCD. QCD predicts that hadrons are composed of smaller hypothetical particles called quarks and gluons. Gluons hold quarks together. QCD says that protons in a nucleus repel each other and that gluons are what hold a nucleus together. This is called the Strong Force.

TGP differs from QCD primarily in the area of gluon particles. TGP states that gluon particles do not exist. The nucleus is a single particle composed of quantum strings. Nuclei have no repulsive force in it. The Strong Force does not exist. Surrounding gravity propulsion forces are what hold all particles together including nuclei. When a nucleus is broken up into multiple particles, some quantum strings radiate as energy.

8.9 Unified Field Theory

The goal of the unified field theory was to explain the following four forces:

---

[15] "Selected Papers on Quantum Electrodynamics", edited by Julian Schwinger (Dover 1958)

1. Electromagnetic force.
2. Weak Force (The force responsible for certain radioactive decay processes.)
3. Strong Force (The force that holds atomic nuclei together)
4. Gravity

The problem with field theory is that it has to explain all forces as fields. Fields are mathematical representations of forces between particles. Fields don't explain how particles interact with each other without touching. They just do. The unified field theory tries to explain the first three forces but cannot explain how gravity works.

TGP differs from previous field theory in the following ways.
1. TGP explains how all of these four basic forces work. TGP's Quantum String Theory explains how gravity works. Previous field theory could not.
2. TGP explains how all particles physically interact with each other when they are apart and not touching. Field theory cannot and never will.
3. TGP further explains how the following additional forces work: inertia forces, centrifugal forces, particle magnetic pole forces, all attractive forces, all repulsion forces, forces that holds all particles together and Lamb shift forces.

Based on TGP, field theory can now develop high-level equations to approximate all fields and forces.

8.10 Superstring Theory

Here is an overview of the superstring theory. All matter is made of tiny vibrating strings. Each string is about 100 billion billion times smaller than a neutron. Strings come in two types, open strings and closed strings. Strings have different modes of vibration. Vibrations are a resonance of each other. Particles are defined by the type of strings they have, open or closed, and their mode of vibration. Strings interact with each other five different ways.[16]

---

[16] Michio Kaku and Jennifer Thompson, "Beyond Einstein, The Cosmic Quest for the Theory of the Universe", New York, Anchor Books, 1995

Here are, what I consider to be the first three important milestones in superstring theory:

1. In the 1960s, Geoffrey Chew of the University of California at Berkeley created the S-matrix theory. The S-matrix theory was based on general S-matrix equations. It mathematically described particle collisions.
2. In 1968, Gabriele Veneziano and Mahiko Suzuki rediscovered the Beta function that Leonhard Euler wrote in the 1800s. The Beta function mathematically explained most of the S-Matrix theory.
3. In 1970, Yoichiro Namb of the University of Chicago showed that quantum strings interacting with each other could explain the Beta function.

Namb's original superstring theory was on the right track to explain how quantum strings work. He took the theory to a physical level and to a more elemental level. The theory was a more elemental level because it described all particles as being composed of quantum strings. This was the major breakthrough. However, Namb's superstring theory was not perfect. It had some mathematical problems that were later solved. But the solutions didn't take the theory deeper into a more elemental level. People added high-level concepts to it including spin, bosons, fermions, fields and forces. The problem is that you cannot apply high-level concepts to create a more elemental theory. It's like trying to make Newton's laws of motion explain everything.

TGP is consistent with the superstring theory in assuming that all particles are made of quantum strings. The superstring theory is good at predicting how particles hitting other particles create subatomic particles. This is called "strong interactions." TGP does not conflict with superstring theory's higher-level concepts of strong particle interactions. Therefore, this book does not review or explain superstring theory of strong particle interactions.

TGP compliments and differs from the current superstring theory in the following ways:

*   TGP is a more elemental theory because it additionally explains how quantum strings create particle properties, fields and forces.
*   TGP gives a deeper understanding of fields. Fields are individual traveling quantum strings.
*   All forces among particles and fields are interactions among quantum strings.

- With TGP, fields are not simply represented as multiple dimensions as the previous superstring theory did. Fields are, unfortunately, more complex interactions of huge numbers of quantum strings.

8.11 Questions

The following questions are about statements made in this section. The answers are in Section 16.

8.1   What are two wrong conclusions of the Michelson-Morley experiment that are not accepted by TGP?

8.2   What experiment was Einstein's Theory of Special Relativity based on?

8.3   What is the maximum speed that energy can travel according to The Theory of Special Relativity and The Theory of Gravity Propulsion?

8.4   Describe how gravity is transmitted according The Theory of General Relativity and The Theory of Gravity Propulsion.

8.5   Describe how electrostatic and magnetic energy is transmitted according The Theory of General Relativity and The Theory of Gravity Propulsion.

8.6   In which theory can an electron orbit travel through the nucleus of an atom: The Quantum Mechanics Theory or The Theory of Gravity Propulsion?

8.7   Did the previous Unified Field Theory explain how gravity works?

8.8   Which is a more elemental theory, The Superstring Theory or The Theory of Gravity Propulsion?

8.9   The Quantum Electrodynamics Theory (QED) was initially developed to explain what?

8.10 How does Quantum Chromodynamics Theory (QCD) and TGP differ in explaining what holds a nuclei together?

9.0 TGP's Quantum String Theory

This section explains the Theory of Gravity Propulsion in terms of quantum strings. The previous superstring theory explained how all particles, including all subatomic particles, are created and interact with each other. This section adds to the superstring theory by explaining how quantum strings create all forces including gravity forces, gravity propulsion forces, electrical forces, magnetic forces and strong and weak nuclear forces.

This is a conceptual approach to the Grand Unification Theory that explains all matter and fields. The right conceptual approach must come first then the right math can follow. The supporting math for this theory will be challenging and may be limitless. This is just the beginning. TGP is like a sign that says, "Go this way".

The entire universe, with all of its great diversity, is based on just one type of string. Strings create all matter. A particle is a group of strings. Strings create all energy. Energy is traveling strings. Strings create all forces. Forces are interactions of strings. All kinds of particles, all kinds of energy and all kinds of forces are created by only one kind of quantum string.

Each quantum string has these six properties:
1. Polarity
2. Alignment
3. Frequency
4. Direction
5. Speed
6. Length.

TGP's Quantum String Theory on one level is very simple. You just have strings that bump into other strings. Strings change properties based on their interactions with other strings. You do not have magical and mysterious scientific terms to believe in like "fields", "quantum fields", "time-space", "curvature of time-space", "space-time continuum", "mass energy transformations", "10 dimensional universe" or "zero point energy". You just have strings. Quantum strings can explain all matter, all energies, all forces, all forces, time and space.

9.1 Ether

TGP requires ether to exist. The classical definition of ether is, "Ether is the medium that allows light to travel through space." This is true but vague. Here is TGP's new definition of ether:

"Ether is individual quantum strings traveling through space."

It is these quantum strings traveling through space (ether) that transfers all energy from one particle to another particle over a distance. This is what creates fields. Fields are no longer mysterious and magical forces. The main secret how all forces work is this:

"All particles radiate strings."

These strings travel through space until other particles absorb them. Space is not a vacuum with nothing in it. Space is filled with traveling strings. Space has properties based on the properties of strings that are traveling through it. I call this "ether". If you don't like the term "ether", then you can substitute the words "quantum strings traveling through space". If you understand that fields and energy travel through space, then you know that something is traveling through space. I call this ether.

Ether is all around us. It is the most powerful force in the universe. The power of ether is beyond comprehension. Ether is more powerful than the energy at the center of our Sun. The potential power of ether is over 10^{46}g where g is the force of one earth's gravity. I excluded the derivation of ether power from this book. The point is that the powerful gravity propulsion force, that holds matter together, is only about 10^9g. Ether is over a trillion, trillion, trillion times more powerful than the gravity propulsion force.

9.2 Forces

All distant forces are interactions of particles with traveling strings that were radiated from distant particles. All forces of attraction, repulsion and neutrality can be explained as finite quantum interactions of strings. All forces are due to the amount of interaction and non-interaction among strings from all directions.

Attraction Forces: Two particles attract each other when traveling strings push the particles together more than traveling strings push them apart. In this case, A > B in Figure 28.

Repulsion Forces: Two particles repel each other when traveling strings push them apart more than traveling strings push them together. In this case, B > A in Figure 28.

Neutrality: A particle is neutral when its traveling strings push on another particle the same amount that traveling strings push them together. In this case, A = B in Figure 28.

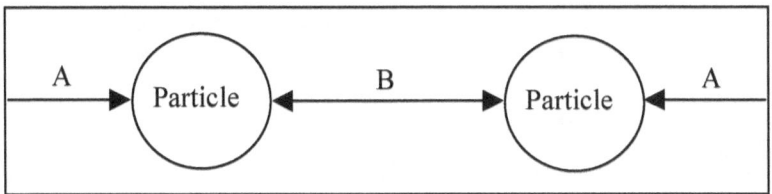

FIG. 28. General diagram of attraction and repulsion forces between two particles.

Which is easier to understand? "Strings can push on particles." Or this? "Magical fields travel through empty space that can mysteriously attract or repel particles." I no longer believe that magical and mysterious "fields" are the final answer. Fields are only an abstract mathematical concept to estimate average effects.

9.3 TGP's Quantum Gravity Theory

The Quantum Gravity Theory is a subset of TGP that describes how quantum strings create gravity. Previously we explained how gravity works in terms of fields. In section 3.1.3, we stated "Mass fields push atoms in a mass outward less than incoming gravity propulsion fields push inward on the same mass. The resultant force is what we call gravity." Now we will explain how gravity works in terms of quantum strings. Mass fields and gravity propulsion fields are composed of traveling quantum strings. What makes these fields different is their general string alignment to each other and their direction of travel.

Gravity depends on the alignment property of all quantum strings. Gravity depends upon how strings are aligned to each other. Gravity is a resultant force of quantum string interactions. TGP concludes that there are no gravity particles, no graviton particles, no quantum gravity strings nor anti-gravity forces.

String alignment is the key to understand gravity. **The more that strings are aligned with (parallel to) each other, the less they interfere (push) each other.** This is from Principle 4C1 which is covered in the next section. String alignment is relative to strings in a mass. Strings in a mass have a general alignment to each other. Mass field strings are partially aligned to strings of the mass. In comparison, alignment of gravity propulsion strings is more randomized. Mass field strings are more aligned to strings in a mass than gravity propulsion field strings are. Because mass field strings are more aligned to mass strings, they push outward less. Because gravity propulsion field strings are more randomly aligned, they push inward more. The difference is gravity.

The table below shows the differences between mass fields and gravity propulsion fields.

	Mass Fields	**Gravity Propulsion Fields**
Direction:	Travels outward from a mass	Travels inward to a mass
Alignment:	Traveling quantum strings are more aligned to strings in a mass	Traveling quantum strings are more randomly aligned to strings in a mass
Force:	Pushes less on a mass	Pushes more on a mass
Gravity:	Pushes outward less	Pushes inward more

The above gravity field table can be explained this way. A mass field is composed of traveling quantum strings. These strings travel outward from the mass. These strings are partially aligned to the strings in the mass. Because of this alignment, they push on the mass less. Therefore, mass field strings push outward less than they would if they were not aligned.

A gravity propulsion field is composed of traveling quantum strings. Some of these strings travel inward (downward) to the mass. These strings are more randomly aligned with the mass strings. Because of this lack of alignment, they push on the mass more. Therefore, gravity propulsion strings push inward more (than mass field strings).

Here is an example of string alignment. Lets say we have a traveling string. The string is aligned at a right angle to its travel. Next we have a group of strings that are standing still. These strings are aligned (parallel) with the traveling string. If you visualize looking down on the top of the strings, they all look like dots. The traveling string is now like a dot going through other dots. It's hard for a dot to hit other dots. And even if the traveling dot met another dot, it only has to push it away the width of the dot in order to travel on. It doesn't have to merge or interfere with it or push it much. It will continue to travel on in nearly the same direction. So basically, the more that strings are aligned with each other, the less they will interfere and push each other.

Strings actually rotate end over end like a cartwheel. It would be more accurate to say, "string rotation angle to its direction of travel." However, I prefer to use the words "tilt" or "alignment".

Strings in a mass try to align to the mass's outward stream of strings. (Principle 4C4 in section 9.4.4) The following explains how strings do this alignment. Outward traveling strings push strings in particles that are not already aligned with it. This process continues until particle strings are aligned with the outward traveling strings. Strings in the mass have plenty of time to align with the outward stream of strings. I will explain the details of this process soon.

The number of inward traveling strings is about the same amount of outward traveling strings. The inward traveling strings are the gravity propulsion strings. Gravity propulsion strings are strings that are not aligned to the radiated mass strings. In general, inward traveling strings are randomly aligned. Strings in the mass cannot align to randomly aligned strings. Therefore the random inward strings (gravity propulsion fields) will push stronger inward on the mass's strings.

Outward traveling strings push against particle strings less than inward traveling strings do. This is because some of the mass strings are aligned to the outward traveling strings. The difference between the weaker outward push and the stronger inward push is gravity.

The next point to explain is how strings align up. If electrons did not orbit and nuclei did not spin, then all strings in a mass would strongly align to each other. Then there would be very little outward push. There would primarily be the tremendous inward push of gravity propulsion strings. The mass would then be a particle of pure mass called a dark hole.

87

Fortunately for us, our electrons orbit and nuclei spin. Because electrons orbit around the nucleus, and because nuclei spin, strings in a particle can only partially align to any pattern of incoming strings that they encounter. To understand how strings line up to mass fields, we will examine the atom's nuclei.

Here are ways that nuclei in a mass can align to the mass field.
1. Nuclei spin alignment: All nuclei spin at the same rate. All nuclei spin in the same direction along a mass' radial line. Note: Nuclei spin rate is not related to its electron orbital rate.

2. Nuclei axis alignment: Nuclei axis tends to align to the average alignments of all incoming strings that the mass radiated.

3. Nuclei string alignment: Strings in the nucleus tend to align (be parallel) to each other and tend to align to the average alignments of all incoming mass strings. The tended string alignment is at an angle to the nuclei axis.

Here are examples of perfect nuclei string alignments to strings radiated outward along the radius. Each example defines the properties of a row of atoms along a radius in a mass. In these examples, atoms radiate strings outward along this radius.

Example: 90-Degree Radiated Strings
1. All nuclei axis are at right angles to the mass radius and are aligned to each other.
2. All nuclei strings are aligned to each other and are aligned to the nuclei axis.

Therefore, all radiated strings are aligned with all nuclear strings as they travel outward along the radius.

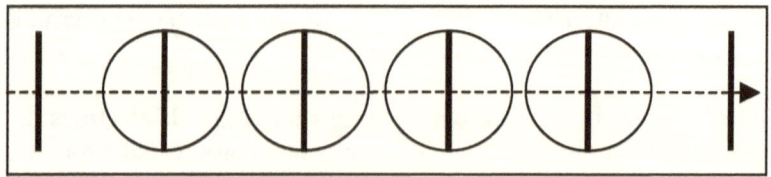

FIG. 29. Example of quantum strings aligned 90^0 to their travel and aligned to each other.

Example: 0-Degree Radiated Strings
1. All nuclei axis are aligned to the radius.
2. All nuclei strings are aligned to each other and are aligned to the nuclei axis.

Therefore, all radiated strings are aligned with all nuclear strings as they travel outward along the radius.

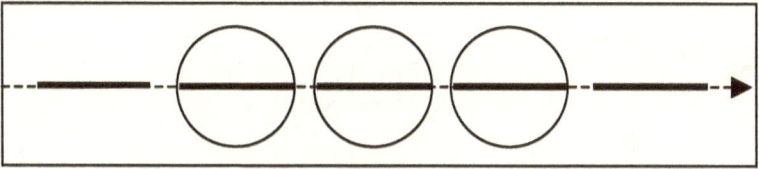

FIG. 30. Example of quantum strings aligned 0^0 to their travel and aligned to each other.

Example: 45-Degree Rotating Radiated Strings
1. All nuclei axis are at right angles to the radius and are aligned to each other.
2. All nuclei spin in the same direction and at the same rate. The spin is synchronized to the radiated mass strings.
3. All nuclei strings are aligned to each other and aligned to a line that is at a 45-degree angle to the middle of the axis.

Therefore, all radiated strings are aligned with all nuclear strings as they travel outward along the radius.

Here is a way to visualize nuclei spin and string alignment in this more complex example. Hold a pen in the middle between your thumb and forefinger. Hold the middle of the pen still and use your other hand to move the bottom of the pen to draw circles. The pen represents how the strings are aligned and moving in the nucleus. Imagine the pen is inside a ball. The diameter of the ball is the length of the pen. The pen just fits inside the ball. The ball represents the nucleus. Now image that the ball is filled with tiny strings that are all parallel to the pen.

Figures 31 and 32 show how traveling strings' tilt will match nuclei strings' tilt in synchronized nucleus. Each circle represents a nucleus. The line in the nuclei represents how all nuclei strings are aligned in parallel to each other. The lines below the circles represent strings traveling from the left to the right that are going through the nuclei. The left side represents the center of a mass that the strings are traveling from. The right side represents the outward side of the mass. Figures 31 and 32 show that the tilt of all nuclei strings exactly matches the tilts of traveling strings. When nuclei radiate strings outward along this line, the strings will have the same tilt as the traveling string. These newly radiated strings are added to the strings that were traveling through them.

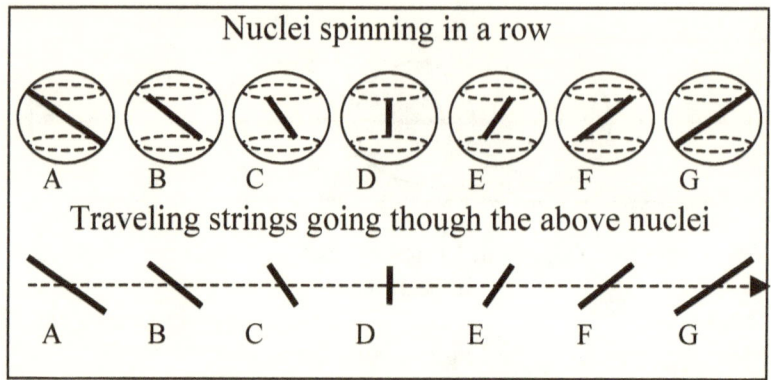

FIG. 31. Example of quantum strings aligned 45^0 to nuclear rotation axis and synchronized to their travel.

Figure 32 shows the traveling strings a moment later as they travel outward. The strings have moved right one position. String position A is now B. Notice that the nuclei spin has changed and the tilt of nuclei strings now match that of the traveling strings as they go through them.

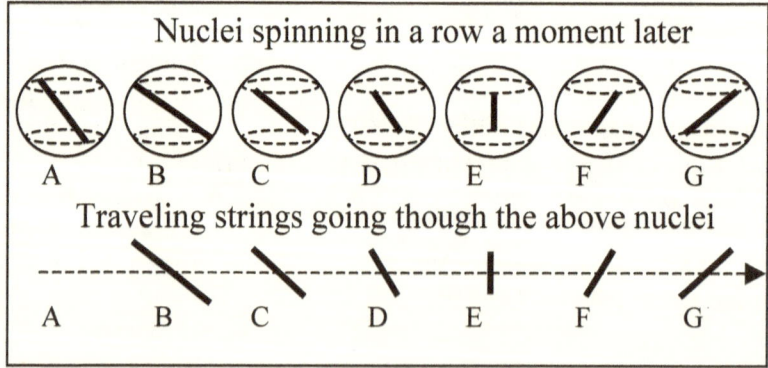

FIG. 32. Example of quantum strings aligned 45^0 to nuclear rotation axis and synchronized to their travel a moment after FIG. 31.

The above examples are an over simplification of how strings align in a mass. We only showed one row of nuclei and one path of radiated strings traveling through them. This was the path from the center of the mass outward. In a real mass, strings radiate from all atoms in all directions. And all atoms get bombarded from strings in all directions. Nuclei will align their axis, spin and strings to the average properties of all incoming radiated mass strings.

These examples show how atoms can align themselves to match the continuous tilt frequencies of radiated mass strings. This is why mass fields push atoms outward less than gravity propulsion fields push atoms inward. The difference is gravity.

9.4 Quantum String Principles

This section explains TGP's principles that govern how quantum forces work. There is only one type of string that creates all particles, energies and forces. All of the diverse particles, energies and forces of the universe are built upon just one type of string. Strings change properties when they interact with other strings.

These principles provide the foundation to understand matter and energy. In this section, the reading becomes more difficult. I tried to present this as simple as I could by using diagrams to describe complex string interactions. I hope everyone that reads this will understand it. If it is any consolation, please keep in mind that it was harder for me to figure this out. Sections 9.4.1 to 9.4.4 are intended for serious researchers to study. If you want, you may skim through this to section 9.5.

9.4.1 Principle 1: Particle Properties

All visible matter is composed of strings that are positive (+Strings) and negative (–Strings). A +String has an opposite polarity of a –String. Here are some examples:
- Electrons are negative. Electrons are composed of more –Strings than +Strings. Thereby they have a net negative polarity (charge).
- Protons are positive. Protons are composed of more +Strings than –Strings. Thereby they have a net positive polarity

(charge). The additional +Strings are approximate to the number of additional -Strings in an electron.
- Neutrons have an equal amount of +Strings and –Strings. Thereby, they have a net neutral polarity (charge).
- Photons are positive. Photons are composed of more +Strings than -Strings.

All anti-matter particles are particles with approximate equal mass and opposite string polarities. Here are some examples:
- Anti-electrons, called positrons, are composed of more +Strings than -Strings.
- Anti-protons are composed of more –Strings than +Strings.
- Anti-photons are composed of more –Strings than +Strings.

9.4.2 Principle 2: Particle Interactions

This section describes TGP's principles how a particle interacts with strings and some other particles.

Principle 2A: Particle Absorption.

When a string meets a particle composed of strings with exact opposite polarities, the particle will absorb the string and be pushed by it. The following are specific examples.

When a +String meets an electron (composed of –Strings), the electron will absorb the +String. The +String will push the electron. The +String will change its polarity, becoming a –String, and become part of the electron.

When a -String meets a proton (composed of +Strings), the proton will absorb the -String. The -String will push the proton. The -String will change its polarity, becoming a +String, and become part of the proton.

There is a possible additional condition to string absorption. The incoming string's frequency times its speed must match the particle's string frequency. This condition means that the particle's string frequencies are maintained.

Principle 2B: Particle Radiation.

All particles radiate strings. When a particle radiates a string, the string changes to the opposite polarity. Here are some examples:

- Electrons (composed of –Strings) radiate +Strings.
- Protons (composed of +Strings) radiate –Strings.
- Neutrons (composed of +Strings and –Strings) radiate equal amounts of +Strings and –Strings.

In space, in general, strings have random polarities.

Principle 2C: Particle Transparency.

If a string or a particle does not have exact opposite polarities of a particle, they will go through the particle and will not affect it.

For example, when a +String meets a proton (composed of +Strings), the string will go right through it and will not affect it. When a -String meets an electron (composed of –Strings), the string will go right through it and will not affect it.

9.4.3 Principle 3: String Properties

TGP has concluded that every string has the following six properties:
1. Polarity (of charge).
2. Alignment.
3. Direction.
4. Frequency.
5. Speed.
6. Length.

Strings are probably closed. This means each string is a loop. Closed strings are required in order for particles to radiate and absorb strings. This theory does not preclude the possibility that strings can alternate between being open and closed. An open string has two ends. This theory requires that strings can be closed and excludes the principle that any type of string is always open in all situations.

The six string properties are described below:

1. **String Polarity.** A string can have any degree and sub-degree of polarity from 0 to 180 degrees relative to another string. For

our point of reference, we define electrons to be composed of strings with 0 degree polarity. A string with a zero degree polarity is called a –String. A string with a 180-degree polarity is called a +String. How strings interact with particles depends on their polarity.

Polarity is a polarity of what we call a "charge", "electric charge" or "electrostatic charge". The net amount of polarity determines the amount of a particle's electrostatic charge. When a particle has more +Strings than –Strings, it has net positive change. When a particle has more –Strings than +Strings, it has a net negative charge. When a particle has an equal amount of – Strings and +Strings, it has a neutral charge.

2. **String Alignment.** Alignment describes how a string is tilted as it travels. A string can have any degree or sub-degree of alignment from 0 to 90 degrees relative to another string's alignment. When strings are aligned to each other, they have a 0 degree alignment and they do not interact with each other. It is string alignment that determines gravity and gravity propulsion forces. I visualize alignment like a pen at a right angle to its direction of travel. It can be turned different ways. How it lines up to other strings is what matters.

3. **String Direction of Travel.** Strings travel in a straight line unless acted upon by other strings. String interactions can change a string's direction of travel. When a string's direction of travel changes, its other properties will change too.

4. **String Frequency.** Frequency is how strings store energy. The higher the frequency, the higher the energy a string has.

5. **String Speed.** Strings can start and stop quickly when they interact with particles. When a string's speed changes, its other properties will change too. From a string's point of view, speed is how fast its frequency travels through itself.

6. **String Length.** All string lengths are the same. String length determines the lowest frequency a string can vibrate at. This is discussed more in the section, "What Created the Big Bang".

9.4.4 Principle 4: String Interactions

String interactions describe what happens when two individual strings meet.

Principle 4A. Conservation of Energy.
String interactions do not loose energy. Their properties change to conserve energy of all strings involved. The conservation of energy equation is:

$$0 = \Delta\text{Polarity} + \Delta\text{Alignment} + \Delta\text{Frequency} + \Delta\text{Speed}$$

Notice that change of direction is not in the equation. This is because direction is not energy. Change of direction is a result of how the other properties change when two strings interfere with each other.

In a second, a single traveling string has trillions of interactions with other strings. These huge amounts of interactions do not necessarily slow the string down.

Principle 4B. String Non-Interference.

Strings with the same alignment do not interfere with other. Individual strings with the same polarity do not interfere with each other. For example, +Strings do not interfere with other +Strings. – Strings do not interfere with other –Strings.

Principle 4C: String Interference.

Individual strings with different polarities and different alignments interfere with each other. The following describes what happens when these two kinds of strings meet.

Principle 4C1: Amount of Interference

The amount that two strings interfere with each other depends on the difference of their polarities and the difference of their alignments. The greater the differences are, the greater their interference will be. The amount of interference is equal to the difference of polarity times the difference of alignment. The equation is:

Interference = ΔPolarity x ΔAlignment

- Interference is a number from 0 to 1 or a percentage from 0 to 100.
- Difference in polarities and alignment are represented as a decimal from 0 to 1.
- When polarities are the same, then ΔPolarity = 0 and there is no interference.
- When polarities are 180 degrees different, then ΔPolarity = 1.
- When alignments are the same, ΔAlignment = 0 and there is no interference.
- When alignments are 90 degrees, then ΔAlignment = 1.

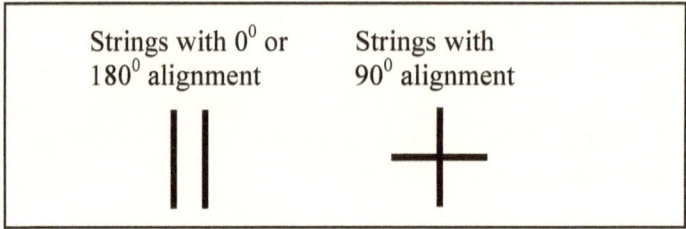

FIG. 33. Examples of quantum string alignments.

Principle 4C2: Change of Direction

String direction changes based on the resultant force of the two strings times the amount of interference. Let's explain these things using diagrams. Figure 34 shows two strings with equal energies (frequencies) impacting each other.

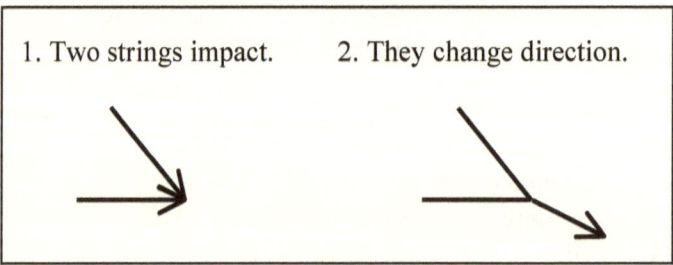

FIG. 34. Example of quantum string impact.

Lets say the two strings (A and B) have different frequencies. String A has three times the frequency of string B. Frequency is energy. We represent energy in vector diagrams by the relative size of the lines. In figure 35 we draw string A with a line three times as long as string B. In this example, lets assume that they interfere with each other 100%.

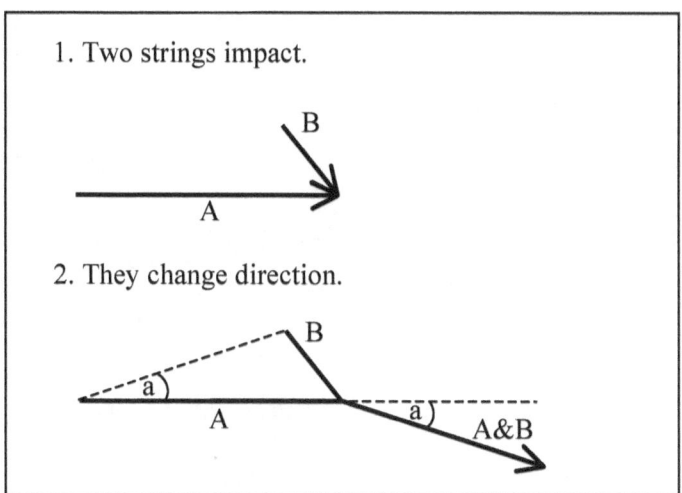

FIG. 35. Example of 100% interference.

In figure 35, the strings change direction. String A is deflected away from the impact at angle "a". String B is deflected to this same resultant path because they both interfere with each other 100%. If you knew the angle of impact, you could calculate angle "a" by using trigonometric functions for a scalene triangle.

Now lets say the strings interfered with each other only 50% instead of 100%. This is how their direction would change.

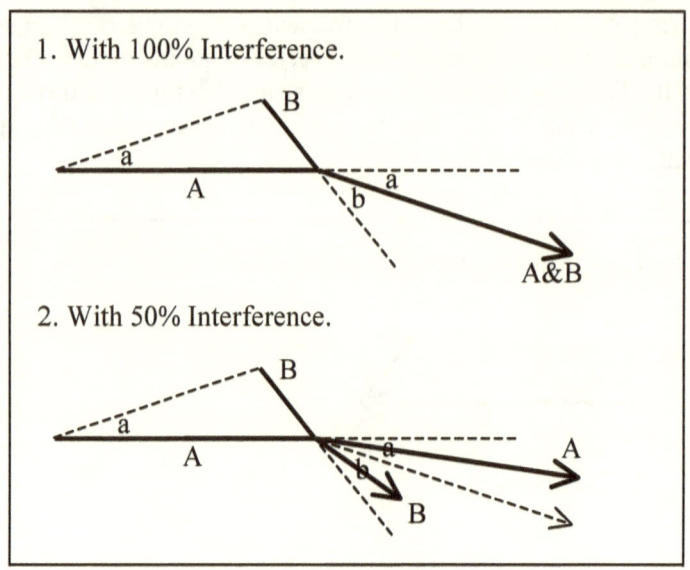

1. With 100% Interference.

2. With 50% Interference.

FIG. 36. Example of 50% string interference.

With 50% interference, string A is deflected only half as much away from its path than it would with 100% interference. It is deflected at an angle of a x 50%. Also string B is deflected half as much away from its path that it would with 100% interference. It is deflected at an angle of b x 50%.

Principle 4C3: Change of Polarity

When strings interfere with each other, their polarities change to some degree. The amounts that string polarities change depends on the following:
- Angle of impact between the two strings.
- Difference in polarities between the two strings.
- Difference of alignment between the two strings.
- Differences in their frequencies.

Polarities don't change when:
- Polarities are the same OR
- Strings are aligned to each other.

In either of these cases, polarities don't change because there is no interference.

Polarities reverse (180 degrees) when:
- The two strings have a 180^0 polarity difference (e.g. −String and +String) AND
- They meet head on (at a zero angle) AND
- They have 90-degree alignment to each other (100% interference) AND
- Their frequencies are the same (equal energies).

In this case, the strings will continue to travel in the same direction and their polarities will be reversed. A −String becomes a +String and visa versa as illustrated in Figure 37.

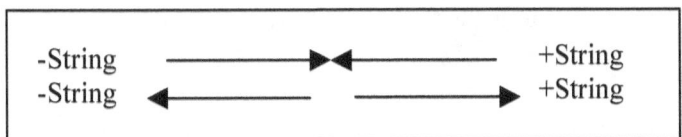

FIG. 37. Example of strings changing polarities with 100% interference.

For all other conditions, the amount of polarity changes somewhere in between. For example, if the strings had 50% alignment, then their polarities would change by half as much (90^0 instead of 180^0). If they meet at a different angle, total change of polarity would be less. If a string had a higher frequency, its polarity would change less and the other string's polarity would change more.

Principle 4C4: Change of Alignment

String alignments change toward 100% alignment proportional to their frequencies and their interference. The general formula is:

New Alignment of String$_1$ = A_1 + (ΔAlignment * $F_1/(F_1 + F_2)$ * ΔPolarity).

New Alignment of String$_2$ = A_2 + (ΔAlignment * $F_2/(F_1 + F_2)$ * ΔPolarity).

Where

A_1 = Alignment of String$_1$.

A_2 = Alignment of String$_2$.

ΔAlignment = Difference in string alignments.

ΔPolarity = Difference in string polarities. This is the amount of interference.

F_1 = String$_1$ frequency.

F_2 = String$_2$ frequency.

Principle 4C5: Change of Speed

Strings dramatically change speeds when particles absorb them and when particles radiate them. Strings can also change speeds when they interfere with other strings traveling at different speeds. Interference averages the speed differences among strings proportional to the amount of interference.

Principle 4C6: Change of Frequency

String frequency is inversely proportional to its speed. When a string slows down, its frequency proportionally increases. When a string speeds up, its frequency proportionally decreases.

The speed that a wave travels through a string is directly proportional to its frequency. As the frequency increases, the wave through the string proportionally increases. As the frequency decreases, the wave through the string proportionally decreases. The table below shows these differences.

String Speed	String Frequency	String Wave Speed
Decreases	Increases	Increases
Increases	Decreases	Decreases

General Interference Diagram

The following Figure 38 shows TGP's general interference diagram for a stream of −Strings and +Strings meeting approximately head on. If they met exactly head on, there would only be a polarity change and no deflection. This diagram shows a stream of −Strings going generally to the right meeting a stream of +Strings going generally to the left. This causes deflections, as an outward cone shape, from their average point of interference. Their polarities are changed. This diagram represents interactions of billions of strings, not just an interaction between two strings. This is a conceptual diagram to illustrate average effects of radiated strings between two particles.

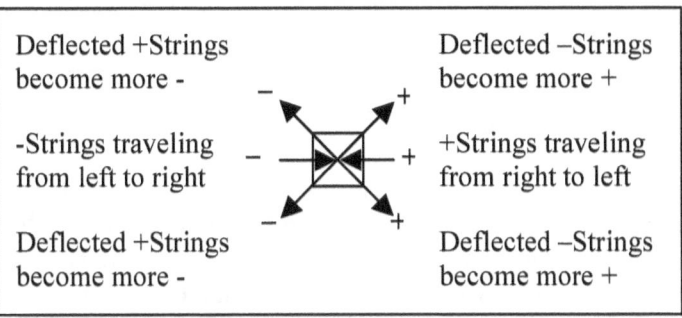

FIG. 38. General Interference Diagram.

Note: I nickname the "General Interference Diagram" simply the "Deflections Diagram". This deflection concept is the key to understand how all attracting and repelling forces work. It is the key to understand the Grand Unification Theory. I worked for some time to figure this out but to no avail. It was driving me crazy. Frustrated, discouraged, depressed and on the advise of my psychologist, I decided to permanently stop working on this and go on to other things. That night, the deflections concept came to me as I slept. Before I went to bed the next night, February 11, 2002, I had completed the TGP's Quantum Electrostatic Theory and all of the particle interference diagrams in this book. I had the biggest headache I ever had that day. And, I was happy.

9.5 Magnetic Poles in Particles

Magnetic poles in particles are formed by symmetric positions of +Strings and −Strings in the particle. +Strings and −Strings are not evenly distributed throughout the particle. The result is that one side of the particle has more of a positive polarity (charge) and the other side has more of negative polarity (charge). This is because strings seek positions of least resistance to the electrostatic forces around them. We call this "magnetic poles" even though it is actually an "electrostatic dipole".

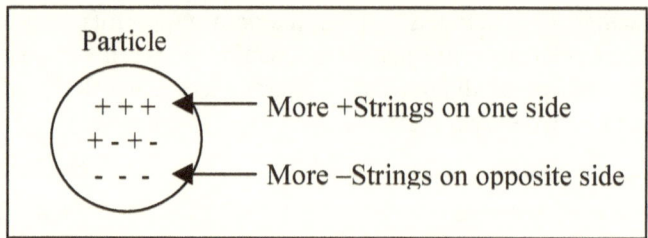

FIG. 39. Particle magnetic poles are electrostatic dipoles.

When we say a particle has "paired strings", it means it has "an equal number of +Strings and –Strings". The strings in a neutron are all paired. This results in a net neutral electrostatic charge (balanced polarity). However, in protons and electrons only some of their strings are paired. The unbalance of paired strings is what gives particles their electrostatic charge (net polarity). A proton has a positive charge because it has more +Strings than –Strings. An electron has a negative charge because it has more – Strings than +Strings.

"Paired strings" does not mean that +Strings and –Strings are physically connected to each other. Strings are free to move around in the particle. Strings in a particle exist together like a compressed gas in a bottle. Paired strings allow particles to form magnetic poles. A magnetic pole is formed when +Strings are on one side and –Strings are on the other side of the particle.

Strings seek positions of least resistance to the electrostatic forces around them. This is similar to the "London Force" which causes molecular dipoles. Orbiting electrons induce poles in the atom's nucleus. Nuclear poles induce poles in orbiting electrons.

9.6 Light (Predictions 13 and 14)

Is light a particle or a wave? It is both. Light is composed of photon particles. Like all particles, photons are composed of quantum strings. Also like all particles, photons radiate strings. The strings that photons radiate have wave properties. There are different ways to describe the same thing: Radiated strings = Mass fields = Wave properties.

The strings in a photon vibrate at a certain rate. The frequency of the photon's radiated strings is close to but is slightly less than the frequency of the photon's internal strings. This is because radiated strings travel slightly faster than the photon. The photon's wave property is close to the photon's property. Here are different ways to describe the same thing:

- Photon particles radiate strings.
- Photons radiate a mass field.
- Photons radiate light waves.
- Light radiates light waves. Or to over simplify,
- Light radiates light.

Note that electrons and protons also have wave properties of their mass fields. Their wave (radiated string) frequencies are greatly lower than their particle string frequencies. This is because of the vast difference in their speeds.

9.6.1 Photons are Positive Particles

In order for electrons to absorb photons, TGP requires photons to have opposite string polarities of electrons. (Principle 2A) Since electrons are composed of –Strings, photons must be composed of +Strings.

Since photons are positive, they should exhibit the same properties of protons except photons are a lot smaller. Electrostatic fields and magnetic fields must affect photons in the same way as they affect protons. There are two interesting ways to prove this: the electrostatic test and the magnetic test.

9.6.2 Photon Electrostatic Test (Prediction 13)

Here is how to test and measure photon's electrostatic properties. Have a laser shoot a light beam between two mirrors. The mirrors reflect the light as many times as possible and still have deflected light be detected. The beam of light travels between two closely spaced high voltage electrostatic plates. Because light is positive, the light should bend toward the negatively charged plate. A detector measures the fringe amount of light on the side with the negatively charged plate. If the light does bend that way, the detector should measure an increased amount of light.

This experiment has some problems to overcome. Any charged molecules or atoms in the light beam would increase air density towards one side of the plates. This would cause refraction. The light should go through

103

a perfect vacuum. I envision the beam going through a dielectric (insulated non-conducting) tube with a small channel in the center for the laser beam. The channel has a vacuum. The tube would have electrostatic plates on both sides of the channel. Another potential problem is any electrons traveling through the tube would push the light toward the positive plate. So, there must be absolutely no current between the two plates. This is not an easy experiment.

9.6.3 Photon Magnetic Test (Prediction 14)

Here is a way to test and measure photons' magnetic properties. Measure how much a star's light is bent around the sun at different positions. The sun has a strong magnetic field. The sun's magnetic field is about 2000 times the strength of earth's magnetic field. It should have a noticeable effect on light.

One problem is that the Sun's magnetic field strength and poles change. Magnetic poles change polarity about every 11 years. Sunspots create more intense magnetic fields than other areas. Its important to know the sun's magnetic pole orientation and strength at the time that light bending is measured. The best effect is when the sun's magnetic poles are at a right angle to our view. This happens once every 6 months.

Another problem is the sun's corona. Sun's gases refract (bend) light. Refraction bends light differently based on its frequency (color). The lower the light frequency is (redder), the more corona gas bends it. So measurements should be made using multiple light frequencies (high/blue and low/red). We can measure how much corona gas bends light from how much blue and red light is bent differently.

Another problems are coordination and timing. This experiment requires coordination of simultaneous precision measurements together with the right conditions. The simultaneous events are listed below:
- One star is near one of the sun's magnetic pole and near the sun's corona.
- Another star is midway between the magnetic poles and near the sun's corona.
- It is desirable to measure more stars in different positions. This dramatically reduces margin of error.
- Sun's magnetic poles are at a right angle to our view. This happens twice a year as we revolve around the sun.

- Measure the sun's magnetic field intensity along the light path of the measured stars.
- Measure how these two stars' light bends using the same multiple light frequencies (high/blue filter and low/red filter).

We have the technology to do this. We have measured and mapped the sun's magnetic fields and intensities. We have measured how starlight bends around the sun. But, we have never done these things simultaneously this way before.

Figure 40 shows how light is bent around the sun at various positions relative to the sun's magnetic field. Light is coming towards you.

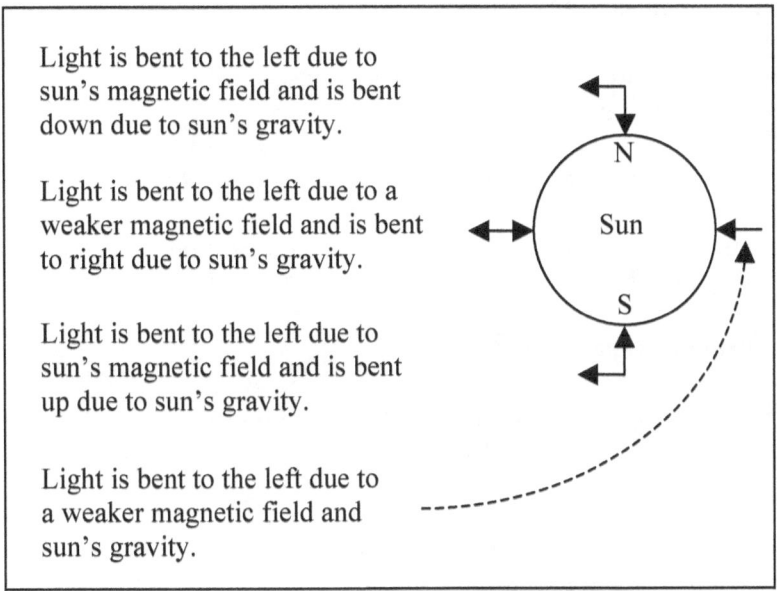

FIG. 40. Sun's magnetic field bends light.

9.7 TGP's Quantum Electrostatic Theory

TGP's Quantum Electrostatic Theory explains how quantum strings create electrostatic fields and forces. "Electrostatic" primarily refers to fields and effects of electrons and protons. We will explain that electrostatic forces are a result of string deflections and changes to string polarities. This section 9.7.X is for serious researchers. If you want, you may skim through this to section 9.8.

Traveling strings create all fields. Strings interactions cause the following established electrostatic forces:
- Electrons repel electrons.
- Protons repel protons.
- Protons and electrons attract each other.
- Neutrons do not attract or repel electrons.
- Neutrons do not attract or repel protons.
- Neutrons do not attract or repel other neutrons.

The following sections explain the basic concept of electrostatic attraction, repulsion and neutrality. To try to keep things simple, the explanations make the following assumptions:
- Protons are composed only of +Strings.
- Electrons are composed only of –Strings.

Actually, protons and electrons also have pairs of +Strings and –Strings. String pairs result in neutral net forces, which is explained later. String pairs cause magnetic poles in particles as explained in section 9.5. We removed magnetic pole effects in order to just explain general electrostatic effects.

9.7.1 Protons Repel Protons

Protons repel each other because the strings they radiate push each other away more than other strings push them together.

Because all particles radiate strings, (Principle 2B) strings encounter various interferences as they travel toward a particle. Figure 41 below shows a proton radiating –Strings to its left. It also shows an equal amount of –Strings and +Strings coming towards the proton from the left to the right. You should imagine that these strings are on the same path. They are shown above and below each other to show their separate interference diagrams.

If we didn't have this interference and their deflections, all incoming – Strings would simply push the proton head on. Because of these deflections, most -Strings pushes the proton at an angle rather than head on. See interference diagram B in Figure 41 below. If these strings hit the proton head on, then they would push the proton more. Because they hit the proton at an angle, the resultant push (to the right) is less. In other words, ◁ pushes less than ⇒ would.

In Figure 41, interference diagram A shows incoming -Strings changing into deflected +Strings. These +Strings do not push on the proton because +Strings do not interfere with other +Strings. (Principle 2C)

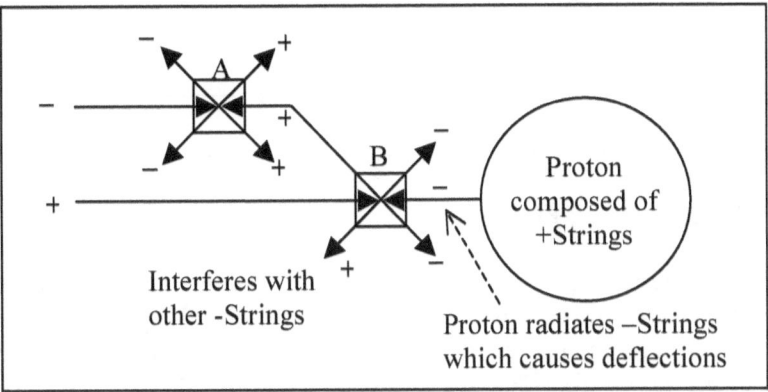

FIG. 41. Proton deflects incoming strings.

In general, a proton radiates the same about of –Strings as it absorbs. The resultant effect is that protons radiate strings outward along their radial lines. In other words, protons radiate strings outward and not at an angle. Figure 42 below shows the radiated –Strings push the other proton outward stronger than other –Strings push the protons together. This is because more of the inward –Strings hit the proton at an angle than the outward radiated strings do. Radiated strings do not interfere with each other because they are the same polarity. (Principle 2C)

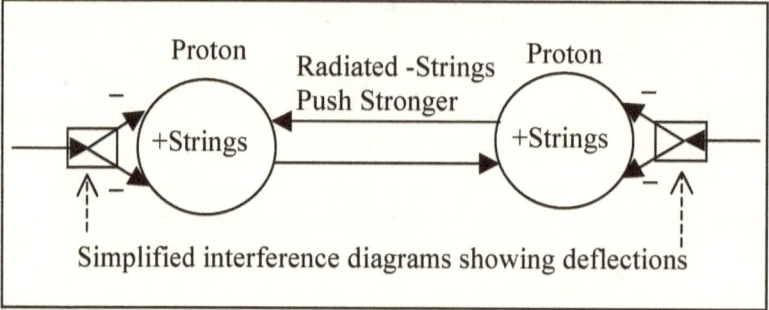

Simplified interference diagrams showing deflections

FIG. 42. Protons repel other protons.

9.7.2 Electrons Repel Electrons

Electrons repel each other for the same reasons that protons repel each other. Its because the strings they radiate push each other away more than other strings push them together. The only difference is that their string polarities are reversed. Electrons are composed of –Strings. Protons are composed of +Strings. The following explanation of electron repulsion is the same as for protons as described above but with their polarities reversed.

Because all particles radiate strings, strings encounter various interferences as they travel toward a particle. Figure 43 below shows an electron radiating +Strings to its left. It also shows an equal amount of – Strings and +Strings coming towards the electron from the left to the right. You should imagine that these strings are on the same path. They are shown above and below each other to show their separate interference diagrams.

If we didn't have this interference and their deflections, all incoming +Strings would simply push the electron head on. Because of these deflections, most +Strings push the electron at an angle rather than head on. See interference diagram B below. If these strings hit the electron head on, then they would push the electron more. Because they hit the electron at an angle, the resultant push (to the right) is less. In other words, pushes less than would.

Interference diagram A shows incoming +Strings changing into deflected –Strings. These –Strings do not push on the electron because – Strings do not interfere with other –Strings. (Principle 2C) It is interference B that pushes the electron to the right.

108

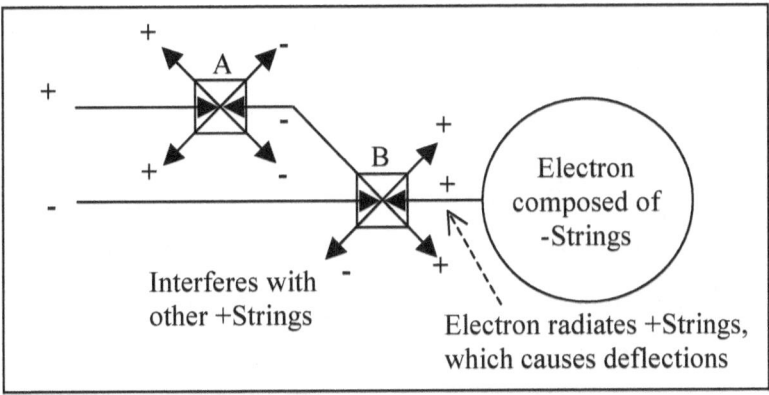

FIG. 43. Electron deflects incoming strings.

An electron radiates the same about of +Strings as it absorbs. Electrons radiate strings outward along their radial lines. In other words, electrons radiate strings directly outward and not at an angle. Figure 44 shows the radiated +Strings push the other electron outward stronger than other +Strings push the protons together. This is because more of the inward +Strings hit the electron at an angle than the outward radiated strings do. Radiated strings do not interfere with each other because they are the same polarity. (Principle 2C)

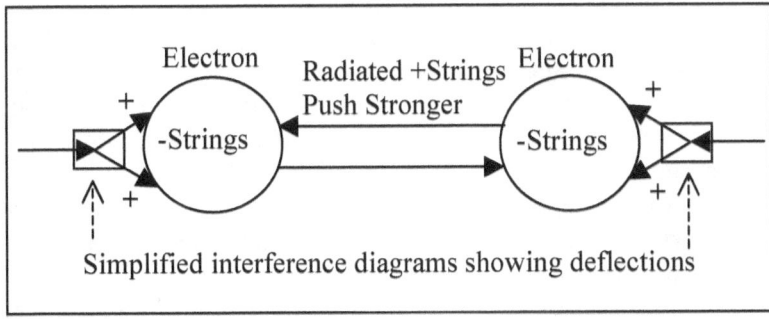

FIG. 44. Electron repels other electron.

9.7.3 Electrons and Protons Attract

Electrons and protons attract each other because strings push them together more than strings push them away. The result is that strings push protons and electrons towards each other. We showed previously how equal amounts of +Strings and −Strings push on protons and electrons. Figure 45 below shows equal amounts of +Strings and −Strings coming from the left and right pushing a proton and an electron together.

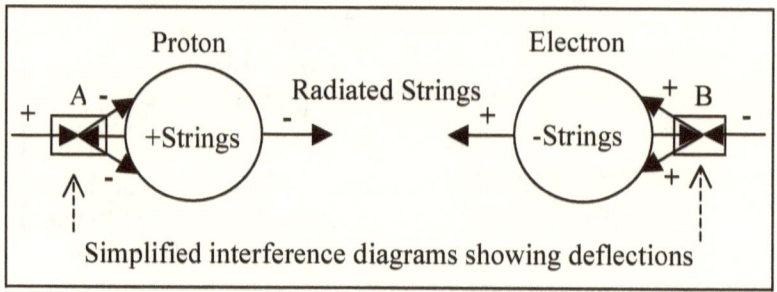

FIG. 45. Proton and electron deflect incoming strings.

Now lets look at the forces that are and are not pushing them the particles apart. Protons and electrons radiate strings. Protons radiate −Strings. Electrons radiate +Strings. If there were no interference, these radiated strings would not push these particles apart. The electron's radiated +Strings would just go through the proton and not push it. This is because a proton is composed of +Strings and radiated +Strings do not interfere with the other +Strings in the proton. (Principle 2C) The proton's radiated −Strings would just go through the electron and not push it. This is because an electron is composed of −Strings and radiated −Strings do not interfere with the other −Strings in the electron. (Principle 2C)

Traveling -Strings and +Strings do interfere with each other as shown by interference diagram C below. This interference does not result in pushing the particles away either. Average interference slightly changes string polarities. Strings are only absorbed by a particle when their polarities are exactly opposite of the particle string's polarities. (Principle 2A) The interference changes the string polarities enough so they still don't push against the other particle.

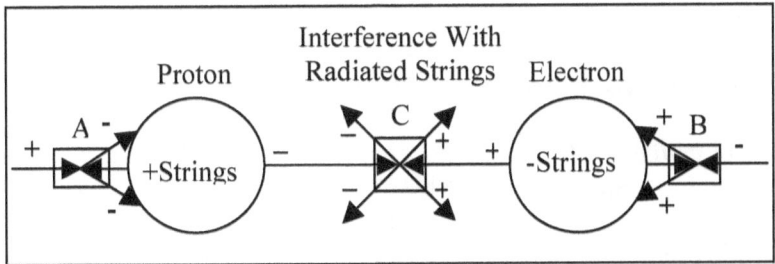

FIG. 46. Radiated string interference in between proton and electron.

Some +Strings (Figure 47, #2) go through the proton without interference. If there were no other interferences, some of these strings would push the electron away. But some of these +Strings interfere with -Strings at D below. This modifies their polarities so they do not push against the electron.

Some –Strings (Figure 47, #4) go through the electron without interference. If there were no other interferences, these strings would push the proton away. But some of these -Strings interfere with +Strings at E below. This modifies their polarities so they do not push against the proton.

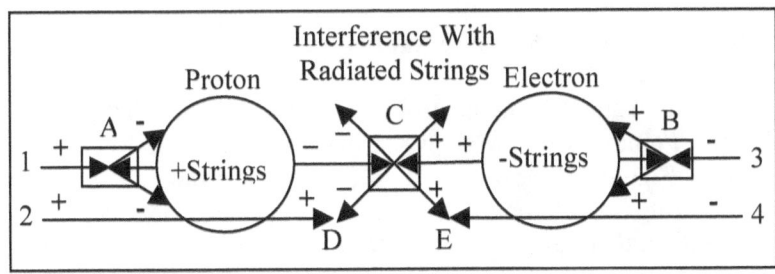

FIG. 47. Protons and electrons attract each other.

We have showed how radiated strings and their interferences with other strings reduce the amount that the two particles are pushed apart. The result is that strings push the proton and electron together.

TGP's fundamental factor is that radiated strings from protons and electrons do not push each other away. Interferences among –Strings and +Strings add complexities of polarity changes and deflections. You can calculate and integrate thousands of string interactions. The bottom line is that the resultant forces are the same.

9.7.4 Neutrons Are Neutral

The following explains how neutrons are electrically neutral. Neutrons do not affect and are not affected by electrons, protons and other neutrons. Neutrons have the following properties:

1. Neutrons are composed of an equal number of +Strings and – Strings.
2. Neutrons absorb and radiate equal amounts of +Strings and – Strings.
3. Neutrons absorb and radiate the same amount of strings per area as electrons and protons do.

Although neutrons are composed of equal amounts of +Strings and – Strings, they are not always evenly distributed at all moments. Dipole effects can happen as strings orient themselves to the fields around them. At one moment in time, there may be more –Strings on one side of the neutron than +Strings.

We will use these properties to explain string interactions that neutrons have with protons, electrons and other neutrons.

9.7.5 Neutrons and Protons Are Neutral

This section explains why neutrons and protons do not electrically affect each other. Figure 48 below shows string interactions between a neutron and proton. Figure 48 show that the neutron and proton absorb the same amount of strings on all sides. Therefore they remain unaffected by each other. The diagram does not show deflections. Deflections are equal on all sides because there are equal amounts of opposing –Strings and +Strings on all sides.

On the average, for every two –Strings that a **proton** absorbs from a direction (6), a **neutron** also absorbs a +String (5) and a –String (7) from the same direction. On the average, for every two –Strings that a **proton** radiates from an area (7), a **neutron** radiates a +String (3) and a –String (4) from an equivalent area. Another way to say this is, a **neutron** radiates (or absorbs) the same amount of strings as the **proton** radiates (or absorbs). Therefore, their net effect is neutral.

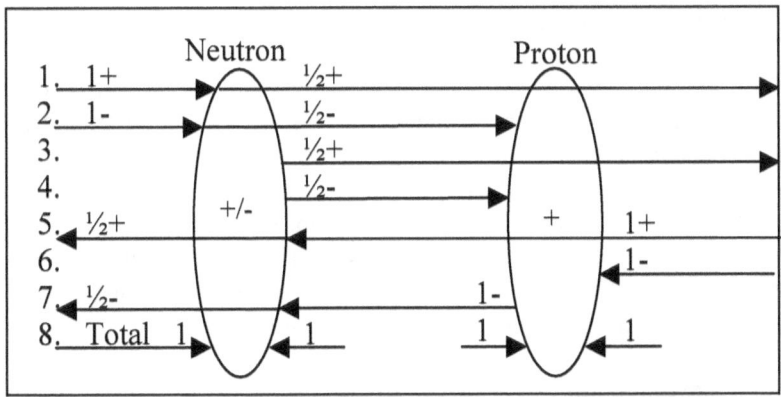

FIG. 48. Neutrons and protons have neutral effects on each other.

The following explains the string interactions numbered in Figure 48.

1. This represents the +Strings coming from space to the neutron. This is the same amount of +Strings that come from space to the proton (5). The neutron absorbs half of these strings. The rest of the +Strings travel on through the proton unaffected.

2. This represents the -Strings coming from space to the neutron. This is the same amount of -Strings that comes from space to the proton (6). The neutron absorbs half of these strings. The rest of the – Strings travel on to be absorbed by the proton.

3. One half of the strings that the neutron radiates are +Strings. These travel on through the proton unaffected.

4. One half of the strings that the neutron radiates are -Strings. The proton absorbs these.

5. This represents the +Strings coming from space to the proton. This is the same amount of +Strings that come from space to the neutron (1). These travel through the proton. The neutron absorbs half of these strings. The rest of the +Strings travel on through the neutron unaffected.

6. This represents the -Strings coming from space. This is the same amount of -Strings that comes from space to the neutron (2). The proton absorbs these.

7. The proton radiates –Strings. The neutron absorbs half of these. The rest of the strings travel on through the neutron.

8. Add up the strings absorbed by the neutron and proton on each of the four sides. They equal one on each side. The net force between neutrons and protons is zero.

9.7.6 Neutrons and Electrons Are Neutral

This section explains why neutrons and electrons do not electrically affect each other. These are is the same reasons why neutron and proton don't affect each other but with different polarities. The following Figure 49 shows string interactions between a neutron and an electron. It shows that the neutron and electron absorb the same amount of strings on all sides. Therefore they remain unaffected by each other. The diagram does not show deflections. Deflections are equal on all sides because there are equal amounts of opposing –Strings and +Strings on all sides.

On the average, for every two +Strings that an electron absorbs from an area (5), a neutron absorbs a +String (7) and a –String (6) from an equivalent area. On the average, for every two +Strings that an electron radiates from an area (7), a neutron radiates a +String (3) and a –String (4) from an equivalent area. Another way to say this is, a neutron radiates (or absorb) the same amount of strings as electrons radiate (or absorb). Therefore, their net effect is neutral.

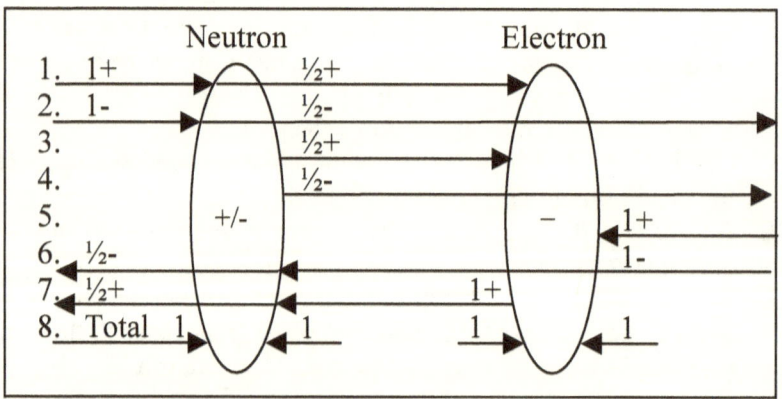

FIG. 49. Neutrons and electrons have neutral effects on each other.

The following explains the string interactions numbered in Figure 49.
1. This represents the +Strings coming from space to the neutron. This is the same amount of +Strings that come from space to the electron (5). The neutron absorbs half of these strings. The rest of the +Strings travel on and the electron absorb them.
2. This represents the -Strings coming from space to the neutron. This is the same amount of -Strings that comes from space to the electron

(6). The neutron absorbs half of these strings. The rest of the −
Strings travel on through the electron.

3. One half of the strings that the neutron radiates are +Strings. The electron absorbs these.

4. One half of the strings that the neutron radiates are -Strings. These travel on through the electron.

5. This represents the +Strings coming from space to the electron. This is the same amount of +Strings that come from space to the neutron (1). The electron absorbs these strings.

6. This represents the -Strings coming from space to the electron. This is the same amount of -Strings that comes from space to the neutron (2). These -Strings travel through the electron. The neutron absorbs half of these. The remaining ½ of the -Strings travel through the neutron.

7. The electron radiates +Strings. The neutron absorbs half of these. The rest of the strings travel on through the neutron.

8. Add up the strings absorbed by the neutron and electron on each side. They all equal one. The net force between neutrons and protons is zero.

9.7.7 Neutrons Do Not Affect Other Neutrons

This section explains why neutrons do not electrically affect other neutrons. Figure 50 below shows string interactions between two neutrons. The diagram shows that the neutrons absorb the same amount of strings on all sides. Therefore they remain unaffected by each other. The diagram does not show deflections. Deflections are equal on all sides because there are equal amounts of opposing −Strings and +Strings on all sides.

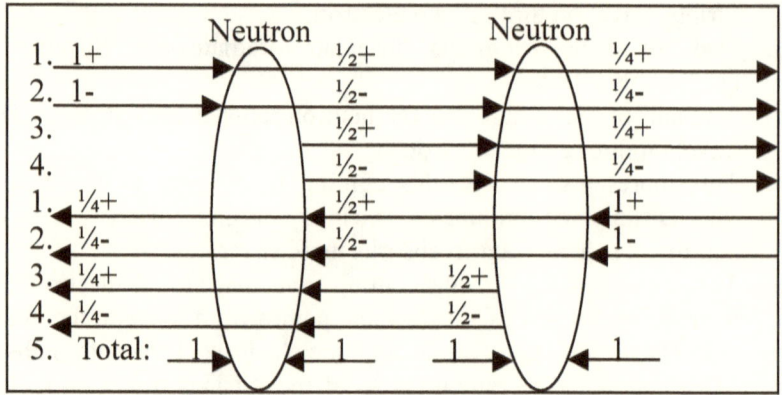

FIG. 50. Neutrons have neutral effects on each other.

The following explains the string interactions numbered in Figure 50.

1. This represents the +Strings coming from space. This is the same amount that impacts the other neutron from space. The neutron absorbs half of these strings. The other half of the +Strings travels on. The other neutron absorbs half of these, which is ¼ of the original +Strings. The rest of the strings travel on, which is ¼ of the original +Strings.

2. This represents the -Strings coming from space. This is the same amount that impacts the other neutron from space. The neutron absorbs half of these strings. The other half of the –Strings travel on. The other neutron absorbs half of these, which is ¼ of the original -Strings. The rest of the strings travel on, which is ¼ of the original -Strings.

3. One half of the strings that the neutron radiates are +Strings. The other neutron absorbs half of these, which is ¼ +Strings. The rest of the strings go on through the neutron.

4. One half of the strings that the neutron radiates are -Strings. The other neutron absorbs half of these, which is ¼ -Strings. The rest of the strings go on through the neutron.

5. Add up the strings absorbed by the neutron and electron on each side. They all equal one. The net force among neutrons is zero.

9.8 String Radiation and Absorption

This section describes TGP's concept how particles radiate strings and absorb strings. First, we will explain how particles radiate strings. Strings vibrate. The vibrations travel up and down the string a little faster than the speed of light. Image that there is a thin stationary wall with small hole in it. The hole is just big enough to fit a string through it. When one part of the string enters the hole, the vibrations in the string push the string against the hole. This pushes the string through the hole. The string is propelled away from the wall at the speed of light because that is the speed of the string's vibrations. The string can have any frequency. What matters is how fast the string's waves travel through the string. That is the speed that it is propelled.

The hole in the wall is probably the center of another closed string. A closed string is a string that has its two ends combined together so it forms into a loop. When a string enters the hole of a closed string, it is propelled through it by its interference with the other string. Both closed strings and open strings can go through the hole of a closed string. An open string becomes a closed string when its two ends meet.

This is how particles absorb strings. Particles absorb traveling strings in reverse of how strings are radiated. A traveling string goes through the center of a closed string in a particle. If the closed string has an exact opposite polarity then the interference stops the incoming string and changes its polarity. The incoming string's frequency times its speed may also need to match the particle's string frequency in order to be captured.

String frequency changes when it is absorbed and radiated. When a particle absorbs a string, the string slows down, its frequency increases and the speed of the wave through the string increases. (Principle 4C6). The string interacts with other strings in the particle. This averages out the string frequencies in the particle. When a particle radiates a string, the string is pushed out by the speed of the waves traveling through it. The string speeds up, its frequency slows down and the speed of the wave through the string slows down (Principle 4C6).

9.9 Nuclear Forces

The primary nuclear forces are called the strong force and the weak force. The strong force is defined as the force that holds the nucleus together. The weak force is the force that makes the nucleus fall apart. The following sections explains TGP's concept how quantum strings create these forces.

9.9.1 The Strong Force

The current definition of the strong force is, "The force that binds together the protons and neutrons in the nucleus." The concept of a strong force was created in order to overcome the theory that there is a repulsive force between protons in the nucleus. However, the strong force does not exist. There is no repulsive force in the nucleus. There are no gluon subatomic particles that hold nuclei together. The nucleus is not a collection of protons and neutrons. The nucleus is a collection of strings that have the potential to become protons and neutrons. The nucleus is a single particle. The strings in the nucleus primarily align with each other and vibrate in harmony with each other. **Strings that are not moving against each other do not repel each other.** This is the same as the strings in an electron. There is no strong force in electrons to keep all the -Strings together. There is no strong force in protons to keep all the +Strings together.

The strong force should be redefined as,

"The strong force is the fusion force required to add neutrons and protons to a nucleus."

This fusion force is required to overcome repulsion among protons as they get closer together and finally to provide string alignment in the nucleus. This force equals additional strings that become part of the nucleus. As a proton approaches a nucleus, other protons in the nucleus repel it. It requires a force to push protons together. This fusion force is equal to additional strings that the incoming proton absorbed. Once the proton is absorbed into the nucleus, all strings align to each other and no longer repel. A nucleus has the potential to create protons, neutrons, photons and other subatomic particles if it is split apart. If the nucleus were split up so it released its strings as stable packets of protons and neutrons, the strings that represent the newly defined strong force would be released as energy.

Physicists were so concerned about what holds the nucleus together, that they didn't care about what holds electrons, protons, neutrons, photons and subatomic particles together. The more important question was, "What holds all particles together?"

The answer is, **"Incoming strings that push on a particle hold the particle together."** Particles are held together because they are being pushed inward on all sides. As stated previously, the strings in a particle do not repel each other. So the particle doesn't need much to keep it together. The particle's radiated strings help to regulate the evenness of incoming strings around it. Figure 51 below shows how this works. The circle represents a cross section of a particle. Each arrow represents a general direction of string travel. The net result is an evenly distributed push inward on the particle.

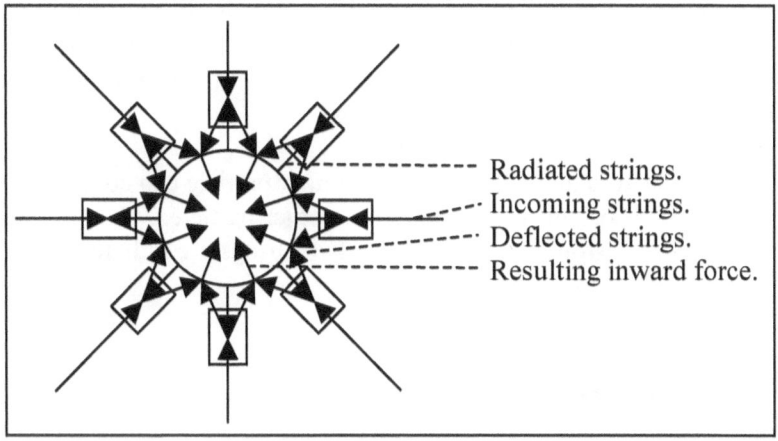

FIG. 51. Incoming strings that push on a particle hold the particle together.

9.9.2 The Weak Force

The current definition of the weak force is, "The force that controls the disintegration of heavy nuclei." A heavy nucleus is like a house of cards. It's just waiting for something to knock it down. If a heavy nucleus encounters an anti-photon, it may be enough to knock it down. It's like a little electron slamming into it. An anti-photon is a photon composed of – Strings which protons absorb. When a proton absorbs a group of –Strings at one time, it becomes less stable.

Nuclei disintegration has to do with stability. Remaining questions are: What makes a particle stable? Why are particles the size that they are?

9.10 Anti-Matter

Anti-matter are particles with equal mass and opposite string polarities. (Section 9.4.1)

Here is an example of an election and its anti-particle, the positron. An election is composed of –Strings. The positron has the same mass of the electron but is composed of +Strings. The electron and positron are attracted to each other the same basic way that electrons and protons attract each other as explained previously.

In a magnetic field, such as that in a cloud chamber, the positron will bend in the opposite direction of the electron and it will bend with same radius. The positron will bend in the same general direction of a proton because they are both composed of +Strings.

When an electron and positron meet, neither particle can completely and simultaneously absorb all the strings of the other particle. The harmony of both particle string groups is disrupted and all strings are radiated away as a burst of energy (as traveling strings).

9.11 Mass Increases With Speed

As a particle travels faster, its mass increases. This has been confirmed experimentally. The only way this could happen is because the particle is moving through a medium that is composed of the same type of mass that the particle is composed of. This medium is ether. Ether is made of the same strings that the particle is made of. Lets look at this in more detail.

As a particle approaches the speed of strings, the strings it radiates forward become a dense cloud ahead of it. This is because these strings are not traveling much faster than the particle is. This dense cloud of strings deflects among themselves. Some strings are deflected back. This increases the rate of strings that the particle absorbs in the forward direction. This increased absorption rate increases the particle's mass. The faster the particle travels, the denser the forward cloud of strings gets and the faster the particle absorbs strings. This steadily increases its mass.

If a particle could go at the speed of strings, the particle could not radiate any strings in the forward direction. The particle would continue to absorb strings faster than it could radiate them as it traveled through the ether. Its mass would continue to grow in size indefinitely.

9.12 Quantum Strings Travel Faster Than Light

If you jumped to this section before reading everything else above, you might be thinking, "Nothing can go faster than the speed of light. I got to check this out first."

Light is made of photons. A photon has energy (E) calculated as hw, where h = Planck's constant and w = frequency. Photons have a mass (M) calculated as hw/C^2 where C = speed of light according to $E=MC^2$ and therefore $M = E/C^2$. A photon is a particle that comes from an electron. When a photon breaks off from an electron, the electron gets smaller and orbits at a lower energy state. When an electron absorbs a photon, the electron gets bigger and orbits at a higher energy state.

According to TGP, photons are made of quantum strings. Quantum strings are smaller than photons. Photons are a big mass compared to quantum strings, like our Sun is bigger than our earth.

It is quantum strings that push photons to give them their speed and energy. No particle can travel faster or as fast as the individual quantum strings that push them. When a quantum string pushes a particle, the particle absorbs the string. The collision speeds up the particle and slows down the quantum string. This is why no particle, including photons, can travel as fast as quantum strings do.

Because photons are so small, there is not much difference between the speed of photons and the speed of quantum strings. Strings travel only a little bit faster than photons do (the speed of light).

Photons radiate strings just like all particles do (Principle 2B). The strings that a photon radiates forward become a dense cloud ahead of it. This is because these strings are not traveling much faster than the photon is. This cloud of strings deflects among themselves. Some strings are deflected back. This increases the rate of strings that the photon absorbs in the forward direction. When the photon absorbs these strings, the strings push the photon back. The faster the photon travels, the denser the forward cloud of strings gets and the more strings push the photon back. Therefore, photons can never travel as fast as individual strings do. This scenario applies to all particles. No particle can travel as fast as individual quantum strings do. This establishes a new speed barrier. TGP establishes a new speed barrier, which is;

"Nothing can accelerate faster than the speed of quantum strings."

The experiment to measure the speed of quantum strings is described in Section 11.3, "Measuring the Speed of Strings".

9.13 Questions

The following questions are about TGP as stated in this section. The answers are in Section 16.

9.1   What is TGP's definition of Ether? Ether is…
9.2   What do all particles radiate?
9.3   What causes an attractive force?
9.4   Electrons have more of what type of quantum string?
9.5   What are the six properties of quantum strings?
9.6   What happens when a traveling –String meets an equal traveling +String head on and they are aligned to each other:
      a) Nothing happens.
      b) They aren't deflected but they change polarities.
      c) They are deflected and their polarities change.
      d) They are deflected but their polarities don't change.
9.7   When a string is deflected, what happens to its polarity?
9.8   What happens when a +String meets an electron?
9.9   What holds particles together?
      a) Gluons.
      b) Incoming strings.
      c) String attractions.
      d) The strong force.
9.10 Anti-matter are particles with equal mass and what kind of string polarity?

10. Wonders of The Universe, Part 2

From the properties of strings, we can understand more wonders of the universe. We have explained how string polarities change, how string polarities interact and how string frequencies change. With these basic properties defined, we can now explain the following wonders of the universe:

- Invisible particles.
- Invisible stars.
- Invisible binary stars.
- Invisible galaxies.
- Invisible galaxy clusters.
- How to detect them.
- What created the Big Bang.

10.1 The Invisible Universe

TGP predicts that there is an invisible universe with invisible mass. Invisible mass does not absorb light. It does not scatter light. Invisible stars radiate invisible light. The world that we see, feel and interact with is composed of +Strings and −Strings (Principle 1). Our matter has string polarities that are 180 degrees out of phase with each other. If a string's polarity is not exactly 0 or 180 degrees then we cannot see or directly interact with it (Principle 2C).

Strings exist with all kinds of polarities. Strings interact with other strings. This changes their polarities to various degrees (Principle 4C3). There are more (invisible) particles with string polarities that are not 0 and 180 degrees, than there are (visible) particles with string polarities that have 0 and 180 degree polarities.

The invisible universe has more mass than the universe that we can see.

Over 99.99% of the universe must be composed of invisible matter. Some people call invisible matter, "dark matter." The term "dark matter" implies that it doesn't radiate light or that it absorbs light. This is not the case. Invisible matter isn't dark. It doesn't absorb our light. Invisible matter is transparent to our light and other of our energy frequencies. Invisible matter can radiate invisible light. I do not use the term "dark matter" for those reasons.

TGP predicts that the invisible universe has:
- Invisible planets.
- Invisible stars.
- Invisible galaxies.
- Invisible light.

For example, lets say there are galaxies composed of particles with string polarities of 1 and 181 degrees. We cannot see these galaxies' light because their photons are 1 degree out of phase with our polarity.

10.2 Invisible Matter Has Gravity

Although we cannot see the invisible universe, TGP predicts we can detect them by their gravity effect. Invisible matter has the same gravity effects as any mass does. It affects visible matter just like it does invisible matter. Therefore, we can detect invisible mass by its gravitational effects on visible matter and light.

A string is a string is a string. Invisible matter radiates strings just like our matter does. All mass radiates individual strings (Principle 2B). Is just that invisible matter starts with strings with a different polarity than ours. These traveling strings interact with other traveling strings. This causes deflections of travel and changes their polarities (Principle 4C3). In a short while, individually traveling strings have random polarities. It doesn't matter if the string came from an invisible mass or visible mass. After a short distance, radiated string polarities from visible and invisible matter become equally randomized. Their gravity effects become equally detectable.

10.3 Halo of Invisible Stars (Prediction 15)

TGP predicts that a halo of invisible stars (HIS) exist around galaxies. They look like they aren't radiating any light or energy, but they are. We just can't see it. TGP predicts that polarity of matter changes as a function of the distance from the center of the galaxy. Therefore, most invisible stars should orbit around and beyond the edges of galaxies. There should be more invisible stars around a galaxy than visible stars in the galaxy.

Verification Status: The galactic halo of invisible stars has gravity that affects the rotation rate of visible stars in the galaxy. If galaxies did not have a halo of invisible stars around them, then visible star's rotation velocities should decrease the farther out they are from the center of the galaxy. This is according to Kepler's law of orbiting bodies, the conservation of angular momentum and the inverse square law of gravity. However, the gravity effect of the outer halo of invisible mass causes inner and outermost stars to rotate around the galaxy closer to the same rate. The difference between in rotation rates determines the amount of invisible mass around a galaxy. For a good explanation of galactic rotations see http://www.owlnet.rice.edu/~spac250/elio/spac.html.

In 1995, Yoshiaki Sofue from the Institute of Astronomy, University of Tokyo, calculated that a halo of dark (or invisible) matter surrounded each galaxy that he measured based on the rotation rate of stars in the galaxy. This dark halo has more than twice the mass of the galaxy. This dark halo has a radius about three times larger than the visible galaxy radius.[17]

There is a project called the MACHO Project to detect the dark matter surrounding our galaxy. MACHO stands for "MAssive Compact Halo Objects". The MACHO Project can detect invisible stars around the rim of our Milky Way galaxy due to their gravitational lens effect. The gravity of an invisible star acts like a lens because its gravity bends light around it. When an invisible star comes exactly between a distant galaxy, and us, the galaxy will increase in brightness for about a week then decrease in brightness over the same amount of time. The shape of these brightness curves are unique to that caused by the gravitational lens effect. The length of time of this brightness curve determines how far away the invisible star is. The amount of the brightening depends on the mass of the invisible star and its distance from us. From this effect we can calculate the mass and distance of the invisible star.

MACHO Project initially assumed (falsely, I argue) that the dark matter they will find would be massive and compact. TGP predicts that the majority of these objects are neither massive nor compact because their

---

[17] "The Most Completely Sampled Rotation Curves For Galaxies", Yoshiaki Sofue, The Astrophysical Journal, 458: 120-131, 2/10/1996. Also see http://adsbit.harvard.edu/cgi-bin/nph-iarticle_query?1996ApJ...458..120S

mass is too small. The galactic halo is not primarily composed of small non-radiating dense bodies like black holes, neutron stars, brown dwarfs or black dwarfs. TGP predicts that the galactic halo is primarily composed of average size stars that are invisible.

10.4 Invisible Galaxy Clusters

John Huchra co-managed the 10-year "CfA Redshift Survey" of 18,000 galaxies. He gave the following statement concerning the results of the survey, "This initial map was quite surprising, showing that the distribution of galaxies in space was anything but random, with galaxies actually appearing to be distributed on surfaces, almost bubble like, surrounding large empty regions, or voids."[18] These voids are huge. The largest is the Boötes Void, which is about 125 million light-years in diameter.

TGP predicts that invisible galaxies fill these voids. If there were no invisible galaxies, then visible galaxies would be pushed by gravity propulsion into the void until they were evenly distributed and there would be no voids. Because there are voids, TGP requires that invisible galaxies fill them.

The matter of invisible galaxies has opposing polarities to each other but they are different polarities than ours. This is why these huge void areas appear empty to us. Matter that is made of the same opposing polarities interacts together. This is similar to the saying, "Birds of a feather flock together."

TGP predicts that mass polarities change as a function of the distance from the center of a void. This is why we see galaxies as walls around voids and why we see galaxies in a line. This is where polarities equal ours so we can see them. Further along the radius, polarities change again so we cannot see the galaxies there. They are there but are invisible. Imagine traveling along a radius of a void at the speed of light. For over 50 million years you see nothing nearby. Then you pass through some galaxies. Then you see nothing again.

---

[18] From http://cfa-www.harvard.edu/~huchra/zcat, 5/5/2002.

10.5 Halo Galaxies

As stated in a previous section, TGP predicts that polarity of matter changes as a function of the distance from the center of the galaxy. Some voids will have invisible galaxies that are close in polarity with us. Some of these invisible galaxies should be surrounded with visible stars. These would appear as a relatively dim halo of stars with a large dark center. These halo galaxies are most likely to appear around the perimeter of a void. Halo galaxies should be very rare and hard to find. But if we do find one, it should be near the edge of a void and there should be others around the edge of that same void at about the same distance from the center of the void.

10.6 The Lamb Shift Effect (Prediction 16)

Invisible stars and galaxies radiate light (photons). There are more invisible photons around us than the light that we can see. We can't see these photons because their polarities are different from ours. We can detect invisible particles by their average electrostatic effect on our atoms. The presence of invisible particles helps to push electron orbits outward. Invisible particles effects inner orbits of electrons stronger than outer electron orbits. We can measure the energy level increase of electron orbits from what they should be. By this method, we can detect the presence of invisible particles including invisible photons (light). Also by this method we can determine the average density and average polarity of invisible particles around us.

I call the effect that invisible particles have on electron orbits, the Lamb Shift. Professor Willis E. Lamb and Professor Polykarp Kusch were the first to discover this effect. In 1947, they discovered that some hydrogen electron orbital states were slightly greater than that predicted by the Dirac theory.[19] The difference is 4.372×10^{-6} eV or 1057.852 Mhz. In 1955, they both received the Nobel Prize in Physics for this discovery. Since then others have verified this and found that the Lamb Effect exists in other atoms too.

The following explains how the Lamb Shift works. First, lets review how protons and electrons attract each other. (Refer to section "Electrons and Protons Attract") Protons radiate –Strings. But only +Strings can push

---

[19] W. E. Lamb, Jr., and R. C. Retherford, "Fine Structure of the Hydrogen Atom by a Microwave Method," Phys. Rev. 72, 241 (1947).

on an electron. Hardly any of the –Strings that protons radiate, get enough interference with other random strings to change their polarity to a +String before they encounter electrons that orbit around them. Therefore, hardly any of the proton's radiated strings push electrons away. This is why electrons and protons attract each other. There are more forces from strings pushing electrons and protons together then there are strings pushing them apart.

Here is how invisible particles change things. When an invisible particle passes inside of electron orbits, then those electrons will be pushed away from the proton more. Radiated strings from the invisible particle interfere with the proton's radiated strings. Where there is an appropriate angle of interference, proton's radiated –Strings will change polarity into +Strings that can now push electrons away. More of the proton's radiated strings change to +Strings than would otherwise happen from interference with random strings. This provides an additional outward push on the electrons. The average electron's orbit will be greater. In addition, the electron orbit will not be smooth. The electron will temporarily move away from the invisible particle when it gets near it.

The presence of invisible particles inside of an electron's orbit causes the following specific effects:
- Increases the electron's average orbit radius.
- Increases electron's average energy state.
- Increases inner electron orbits (e.g. 1P orbital) more than outer orbits.
- The electron obit will have a fluctuating motion that is superimposed on the orbital motion.

Figure 52 below gives an example. This example shows an invisible particle composed of strings with 225-degree polarities and therefore radiates 45-degree strings (180 degree difference). Invisible particles can have any polarity except 0 and 180 degrees. I choose this polarity to keep things simple. When the proton's radiated –Strings meet the invisible particle's radiated 45 degree strings at a 90 degree angle, the interference changes both radiated strings to +Strings. The +Strings can then push the electron away, which slightly increases its orbit.

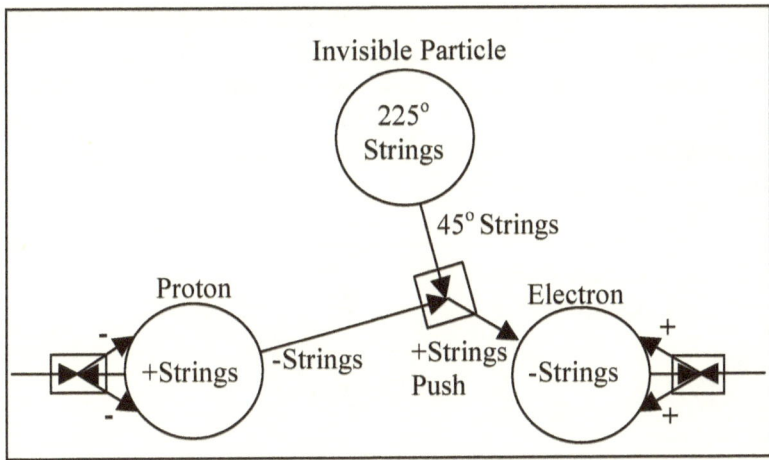

FIG. 52. Lamb Shift: Interference of proton and invisible particle radiated strings push electrons and their orbits outward.

Electrons also push more against the proton. When the electron's radiated +Strings meet the invisible particle's radiated 45-degree strings at a 90-degree angle, the interference changes both radiated strings to -Strings. The -Strings can then push the proton away, which slightly increases its orbit radius.

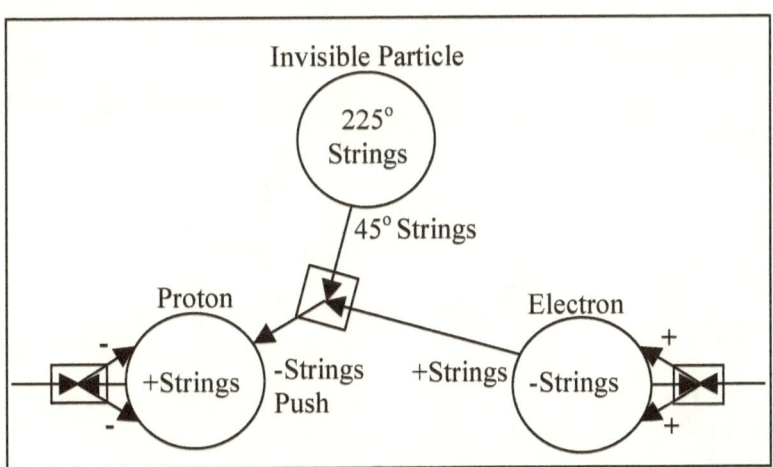

FIG. 53. Lamb Shift: Interference of electron and invisible particle radiated strings push proton away some.

The invisible particle's effect on electrons and protons doesn't just happen at one point. It happens during most of the electron's orbit. This causes a slightly distorted orbit. The following Figure 54 shows the portion of electron orbit that the invisible particle effects. All of the interference angles shown are the same angles. Figure 54 shows interference involving the proton's radiated strings.

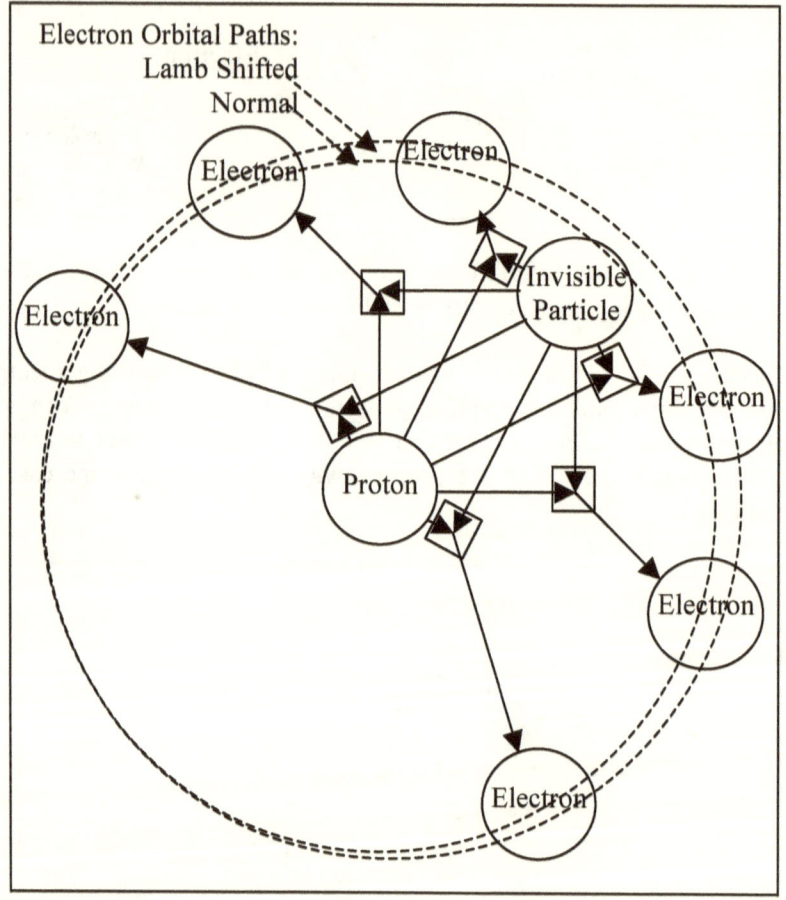

FIG. 54. Lamb Shift: Interference between proton and invisible particle radiated strings repel electron making its orbit larger and wobble.

String interference is more complicated than indicated in the above figures. Along all these paths you have interference from other strings with random polarities. The paths represent high probabilities for an average string polarity and direction of travel.

Verification Status: The Lamb Shift effect has been measured for various atoms. TGP predicts that the density of invisible particles and their average polarity are the same everywhere on earth. With the right particle interaction equations, we can calculate the average density and average polarity of invisible particles around us. By applying this average invisible particle density and polarity to interactions with electrons, should exactly match the various measured Lamb Shift effects. Someone needs to do the math.

10.7 Invisible Binary Stars

TGP predicts a theoretical possibility that there are invisible binary stars. Binary stars are two invisible stars orbiting each other. If they exist at all, they should be extremely rare. There is also a theoretical possibility that we can detect some of them. This requires each invisible star to be invisible to each other and to us. I mention this concept mainly for academic and theoretical discussions.

These stars' orbits can be very small. The orbit radius can be less than their radius of the stars themselves. Their masses don't collide because their polarities don't match and therefore don't interfere with each other.

Interference of their radiated strings causes deflections and changes string polarities. String polarities change based on the angle of impact. At some angle, defected strings might match our polarity. If the intensity of radiated strings is strong enough, fusion (combining) of strings might create photons. If this is possible, then at each half orbit, we might detect these radiated photons as a pulse. The pulse would have photons and anti-photons. Most anti-photons would be annihilated as they traveled into our galaxy.

The following Figure 55 shows two invisible stars orbiting each other. Each star is invisible to each other and to us. This means they have different polarities to each other and to us. They radiate strings and create photons that we can see. These photons radiate from in between the two stars outward.

Larry James

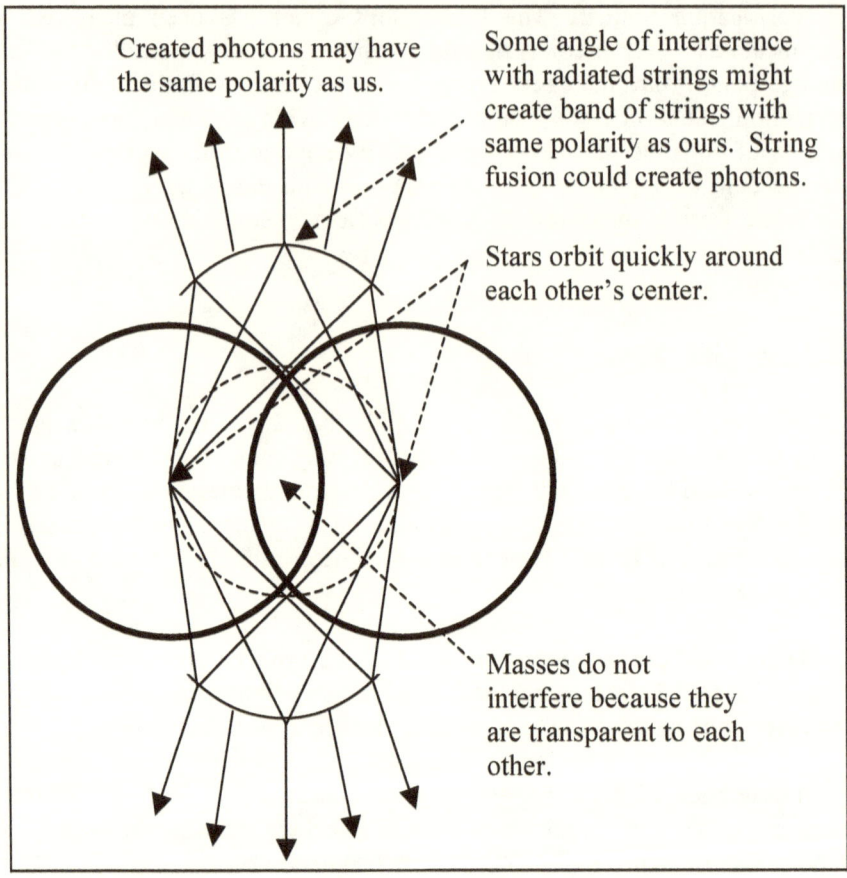

FIG. 55. Theoretical interaction of some invisible binary stars.

As the orbit radius reduces in size, orbital speed increases and the photons, that we might see, radiate in a narrower beam. This narrower beam reduces the overall intensity that we might detect.

The unique trick here is the creation of photons. The two invisible stars radiate strings. Some of those strings may turn into photons. Basically, strings with the same polarity traveling in the same direction, will group together. If there is enough density of strings with the same polarity traveling in the same direction, they might form into a photon. If the stars are large enough, there could be an intense band in between them about the thickness of a photon that creates photons with our polarity.

10.8 What Created The Big Bang

You are probably going to either love or hate this following theory. I won't blame you if you hate it. It's only an overview. If you hate it, then please just think of it as an entertaining idea. This theory is a deduction based on string properties (Principle 3) and string interactions (Principle 4). I have to go where deductions force me to go no matter how strange they seem.

What created the Big Bang? The answer is:

The zero dimension created the Big Bang.

Lets explain this starting at the end of the universe. As we explained in section 7.16, the universe ends for all practical purposes, when all stars burn out. After that, as gravity propulsion forces near zero, all particles come apart. Then there will only be traveling strings interacting with each other. As strings fly apart, they will disperse so much that they stop interacting with each other. When they stop interacting with each other, they stop vibrating. When a· string stops vibrating it has no energy and looses its properties. It returns back to its original single property. A string's original single property is a zero dimension.

The zero dimension is as a single point in space and in time (an instant). When strings revert back to their original property, they exist together at the same point at the same time. At that point, they have no choice but to interfere with each other. It is interference among strings that creates their properties, causes them to vibrate and creates their existence. It is from this point where strings exploded into existence. The zero dimension existed in our universe at one point and at one instant in time. This point was the center of the big bang where all matter and energy started.

Let's look at the zero dimension from its point of view.
1. The zero dimension existed at an instant in time and at a single point.
2. When strings stops vibrating, they exist in the zero dimension. All strings exist there at the same moment in an infinitival point. An infinite number of strings can fit inside a point because they have no width. Strings have width only when they vibrate. Strings can only exist in the zero dimension when they stop vibrating.

133

3. When strings exist in the zero dimension, they interfere with all of the other strings there, which causes them to begin to vibrate. Strings enter the zero dimension from different angles and orientations.

4. When strings vibrate they can no longer exist in the zero dimension. They then pour out of that point at the same moment. This is why the zero dimension can only exist for an instant.

Actually, there is only one string. This string has gone through this cycle near infinity times. A string comes out of the zero dimension, lives, dies and returns to the zero dimension. The cycle continues with the string duplicating itself each time. If you could go back in time, you could go back and see yourself. Each time you go back this way, you are making a copy of yourself at that moment. The quantum string lasts forever. It can go back to its original property an infinite number of times. In other words, it can duplicate itself an infinite number of times.

The length of a cycle to duplicate a string is the life of the universe. The life of the quantum string is the number of current quantum strings times the life of the universe. This is how long the string has lived.

Quantum String Life = Number of Quantum Strings x Life of the Universe.

If the universe required different types of quantum strings then this cycle could not happen. But the universe is built on the same type of string (Principle 3). This string has duplicated itself to create all the matter and energy that we have today. We are in the final steady state of this duplication cycle. In other words, no more strings will be duplicated. We exist in the final steady state. The zero dimension is a one-time event.

I know this is mind boggling because I explained this as a sequence in time. Now you need to understand this from time's point of view. TGP predicts "The Continuity Principle" which states:

The only thing that can exist is the final steady state timeline.

From infinity's point of view, the only thing that can exist is the final steady state. From a time loop point of view, there are no causal loops. It is impossible to create impossible situations. For example, you can't go back in time and kill your parents. There is only the final effect. We are the result of everything that has happened in the past. We are also the result of everything in the future that went back in the past. This includes the life cycle of strings.

This zero dimension theory requires that is possible for a string to stop vibrating. A string might stop vibrating when it no longer interacts with other strings. It stops vibrating when it has its final (last) string interaction. The final string interaction does not have to create a string with a zero frequency. It only has to create a string with a low enough (finite) frequency. All strings have a fixed length (Principle 3). A string cannot support a frequency with a wavelength longer than the string's length. Therefore, it is possible to reach the string's lowest frequency. As a string's speed increases, its frequency decreases. (Principle 4C6) As a string reaches the edge of the universe, its speed increases as strings push each other outward faster. As strings reach their maximum speed, their frequencies nears zero. The final string interaction causes one of the strings to reach its lowest frequency and stop vibrating. This string then becomes a zero dimension and enters the zero dimension.

The universe and eternity are composed of this one string. Excuse me for being "doubly redundant" but in this case I think it is justified.

It is a single, one and only, string that creates all particles, all subatomic particles, all antimatter, all invisible matter, all energies, all fields, all forces, all time and all space.

I now have respect for **THE** quantum string.

10.9 Questions

The following questions are about TGP as stated in this section. The answers are in Section 16.

10.1 Invisible matter is composed of quantum strings with what kind of polarities?
10.2 How can we detect invisible matter?
10.3 What is a halo of invisible stars (HIS)?
10.4 What does the Boötes Void have in it besides space?
10.5 What is a halo galaxy?
10.6 What causes the Lamb Shift?
10.7 What are invisible binary stars?
10.8 What is "The Continuity Principle"?
10.9 What created the Big Bang?
10.10 What do quantum strings create?

11.0 Gravity Communicators (Prediction 17)

We can communicate using gravity fields. Gravity fields travel a little faster than light. Gravity is an alignment property of all strings. Gravity travels at the speed of strings. Therefore, gravity communications are faster than the speed of light.

To transmit a gravity signal, just vibrate a mass. The mass then radiates increased gravity propulsion fields. Varying the intensity of gravity propulsion fields will affect other masses by vibrating them. To receive a signal, we detect how a mass vibrates.

11.1 Gravity Transmitters

There are various ways to have a signal vibrate a mass. One way is to use magnetic fields. Another way is to use electrostatic fields. These gravity transmitters are omni-directional. This means that the signal goes out in all directions. This is unlike radio transmissions. Radio transmitters can use directional antennas to focus radio power in any direction they want. This means that gravity transmitters require a lot more power than radio transmitters do to transmit a signal in a given direction.

There are two types of gravity transmitters: Magnetic and Electrostatic. These are explained below.

11.1.1 Magnetic Gravity Transmitters

A magnetic gravity transmitter uses a magnet suspended inside an electric coil. The magnet is suspended so it is free to vibrate in the coil. Placing an alternating electric signal through the coil vibrates the magnet. On each cycle of the signal, the magnet accelerates, decelerates, accelerates and decelerates. As the magnet accelerates and decelerates, it generates gravity propulsion fields. Changing the signal intensity directly changes the intensity of the generated gravity propulsion field. This is called amplitude modulation (AM). The gravity detector detects changes in gravity propulsion field intensities.

The frequency should be the resonant frequency of the magnet, coil and circuit. This is the frequency that provides maximum output current when only the frequency is changed. This frequency should then stay the same.

Voice frequencies or tones then modulate the amplitude of this resonant frequency. The resonant frequency should be designed to be above 9 KHz in order to provide good voice response.

11.1.2 Electrostatic Gravity Transmitters

An electrostatic gravity transmitter uses electrostatic fields to move one or more plates. The plates are identical in shape to each other. They can conduct electricity. They are flat and round. The edges are rounded to reduce electrostatic discharge. These plates are suspended so their flat sides are facing each other. They placed close together less than an inch apart. They are free to vibrate away and towards each other. Placing alternating opposite voltages on the plates cause them vibrate apart and together. When using air as the medium, the average voltage should be around 4 Kv. On each cycle of the signal, the plates accelerate, decelerate, accelerate and decelerate. As the plates accelerate and decelerate, they generate gravity propulsion fields. Changing the signal voltage directly changes the generated gravity propulsion field intensity. This is called amplitude modulation (AM). The gravity detector detects changes in gravity propulsion fields.

The frequency should be the resonant frequency of the plates and circuit. This is the frequency that provides maximum voltage when only the frequency is changed. Voice frequencies or tones should then modulate the voltage of this resonant frequency.

11.1.3 High Fidelity Gravity Transmissions

Gravity detectors only detect changes in gravity propulsion fields. A steady tone at the same intensity generates the same level of gravity propulsion fields. Continuous tones have to be encoded and decoded. Otherwise, music isn't music and voices sound like mechanical robots. You could use FM (frequency modulation) to modulate the amplitude of the chosen frequency. This is FM on top of AM. To provide high fidelity transmissions, the base resonant frequency should be above 18 KHz.

11.2 Gravity Detectors

There are various ways to create gravity detectors. All gravity detectors should have the following common features:
  1. Detects and amplifies movement of a mass.

137

2. Mass is suspended to have freedom to move.
3. Mass is suspended in a vacuum. This eliminates movement due to sound.
4. Detector is suspended to minimize movement of the earth.

There are two basic types of gravity detectors: magnetic and optical. These are described next.

11.2.1 Magnetic Gravity Detectors

Magnetic gravity detectors use magnetic fields to detect vibrations of a mass. The mass is either a magnet or an iron alloy mass in a magnetic field. Nearby inductor coils detect vibrations of the mass. When the mass vibrates, it induces a current in the nearby inductor coils. Because the mass is conductive, electromagnetic fields can cause it to vibrate. Inductor coils may also pick up electromagnetic fields. For these reasons the detector should be completely enclosed in conducting material to reduce interference from outside electromagnetic fields.

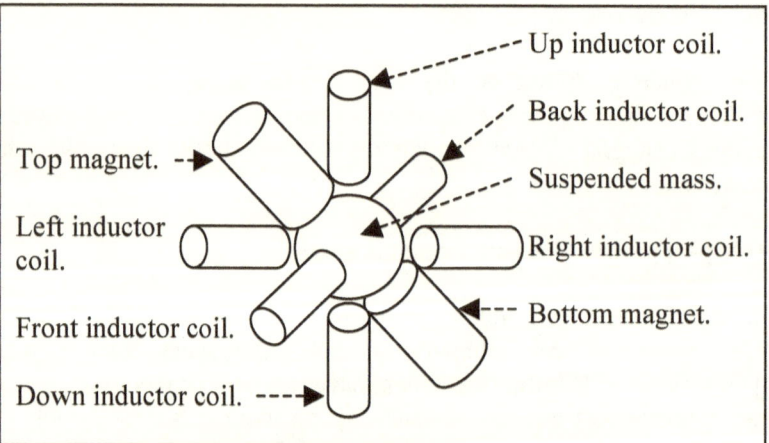

FIG. 56. Magnetic gravity detector.

The up and down inductor coils primarily measure mass vibrations left-right and up-down. The left and right inductor coils primarily measure mass vibrations up-down and front-back. The front and back inductor coils primarily measure mass vibrations left-right and up-down. Put these together with proper circuits and we can focus and amplify gravity propulsion vibrations coming from a specific direction. This is like changing channels on a radio or TV. Instead of changing frequencies, we just look in a specific direction.

You can make an inexpensive gravity detector using just one inductor coil. But you won't be able to focus in a specific direction. You will pick up a lot of background noise.

11.2.2 Optical Gravity Detectors

Optical gravity detectors use laser beams or focused light to detect mass vibrations. Light is reflected off the mass to detectors. As the mass vibrates, light is reflected at different angles. The light detectors then detect the vibrations. The mass material should not be magnetic or conductive. This eliminates interference from electromagnetic fields. The detector does not need to be completely enclosed in conducting material. But it still needs to be enclosed in a vacuum to eliminate vibrations due to sounds. The main advantage of optical detectors is that they can detect minuet vibrations. Also the circuitry is less complex to focus and amplify gravity propulsions fields. With the right signal processing circuitry, you should be able to focus on and pick up gravity waves from particular stars.

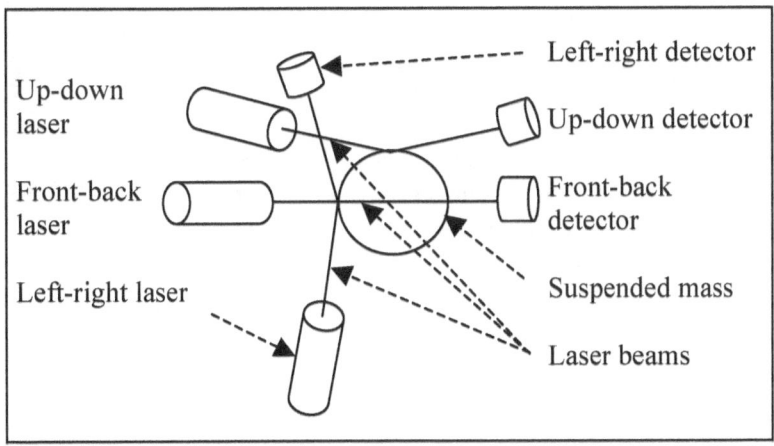

FIG. 57. Optical gravity detector.

The up and down laser primarily measures mass vibrations left-right and up-down. The left and right laser primarily measures mass vibrations up-down and front-back. The front and back laser primarily measures mass vibrations left-right and up-down. Put these together with proper circuits

139

and we can focus and amplify gravity propulsion vibrations coming from a specific direction.

11.3 Measuring the Speed of Strings (Prediction 18)

This section describes how to measure the speed of strings. Because gravity is a property of strings, the speed of gravity and the speed of strings are the same. First you need a gravity transmitter and a gravity detector as described above. Then you measure the difference between the speed of gravity and the speed of light. TGP predicts that the speed of gravity is faster than the speed of light. Here is the general approach to measure the difference in speeds.

1. Get a laser and optical sensor. A pen laser is an inexpensive light source. The idea is to reflect light off of the vibrating gravity transmitter to an optical sensor. The optical sensor should detect the gravity transmitter vibrations. Both the gravity detector and optical sensor should be able to receive the gravity transmitter signal.

2. Equal distance. The distance from the middle of the gravity transmitter to the gravity detector should be the same as the distance from the reflected beam to the optical sensor. Place the optical sensor next to the gravity detector.

3. Create a differential circuit that shows the difference between the optical sensor output and gravity detector output. The differential circuit should be able to show slight differences in a nanosecond. You will need a scope to do this. Adjust it so the output is zero. Any later differences between the optical signal and gravity signal would then show up as a sine wave.

4. Place the gravity transmitter close to the optical sensor and gravity detector. Transmit pulses over the gravity transmitter. Adjust the differential output to zero. This adjusts and compensates for delays caused by the electronics and inertia.

5. Move the gravity transmitter as far away as you can and still detect its signal. The greater the distance, the greater the difference in the two speeds you can detect.

If you then get any differential signal at all, it proves there is a difference in speed between gravity and light. To prove which is faster, you need to know which way the differentiated signal shifted. The amplitude of this differentiated signal shows the difference in speed. The percentage difference in speed is the ratio of the amplitude of the differentiated signal to one of the non-differentiated signals.

One of the problems with this experiment is inertia. Inertia of the receiving mass causes a delay in mass movement and therefore a delay in detecting the signal. For this experiment, the receiving mass should be as small as possible.

11.4 Questions

The following questions are about TGP as stated in this section. The answers are in Section 16.

11.1   How does a gravity transmitter transmit gravity signals?
11.2   What are the two types of gravity transmitters?
11.3   What is a problem that gravity transmitters have that radio transmitters do not have?
11.4   What are the two types of gravity receivers?
11.5   Which travels faster; a signal from a gravity transmitter or a signal from a laser?

12.0 James Names

This book reveals a few new discoveries and predictions. Since they have no names, you can call them James this or that. The good news is that "James" is easier to spell than "Einstein". The following are brief explanations of the James names. They are listed in alphabetical order.

Name	Description
James Anti-matter	Defines anti-matter as particles with approximate equal mass and opposite string polarities.
James Barrier	Nothing can accelerate faster than the speed of quantum strings.
James Communications	The ability to communicate faster than the speed of light.
James Constant	The average gravity propulsion force in Earth's orbit. P_o = 2,121,637,049 approximately.
James Continuity Principle	The only thing that can exist is the final steady state timeline.
James Creation	The universe began from the zero dimension.
James Deflection Effect	Gravity propulsion fields are deflected around a spinning or vibrating object.
James Energy	Individual quantum strings transmit pure energy including electrostatic, magnetic and gravity.
James Entropy	We are losing energy and mass. Our speed of time is slowing down. The universe will end.
James Ether	Ether is individual quantum strings traveling through space.
James Expansion	Gravity propulsion is accelerating our universe's expansion.
James Grand Unification Theory	Same as James' Theory of Gravity Propulsion.
James Gravity Equation	$\Delta G = D_t - D_a - I$
James Gravity Field Theory	The theory that gravity is the difference between a mass field and gravity propulsion fields acting on an object.

James Gravity Fields	Gravity fields are mass fields and gravity propulsion fields.
James Invisible Matter	Matter is invisible when its quantum string polarities are out of phase with our quantum string polarities.
James Lift Effect	The increased upward gravity propulsion force that contributes to lifting a spinning or vibrating object.
James Photons	Photons have a positive electrostatic charge.
James Polarity	Quantum strings have various degrees of polarity of charge. All visible matter has 0 or 180 degrees of quantum polarity.
James Quantum Electrostatic Theory	The theory that polarities and deflections of quantum strings cause electrostatic fields.
James Quantum Gravity Theory	The theory that alignment of quantum strings cause gravity.
James Second Law of Motion	F=MD. Inertia force is the result of gravity field deflections created by accelerating a mass.
James Speed Shift	The apparent variation in effective frequency due to changes in the speed of light between the source and destination.
James Theory of Gravity Propulsion	The theory that a single type of quantum string creates all matter, energy, fields, forces, time and space.
James Time Equation	Time $= 1 + (D_t - D_a + I - G)/P_o$
James Time Shift	The apparent change of frequency due to differences in the speed of time between the source and destination.
James Unified Field Theory	The theory that a single type of quantum string creates all fields, forces and energy.

13.0 Final Question

This is the final overall question:

"Identify the facts presented in this report that support the Theory of Gravity Propulsion. Identify the TGP prediction that each fact supports."

The answers are in Section 16.

14.0 The Web Site

This report is just the beginning. There is a lot of work to be done. The problem is as soon as this report is published, it won't be up to date. You may need reliable and up to date information in these fields to stay valuable in your profession. To help you stay current and to help advance the theories in this report, there is a web site called www.universalpower.org.

This web site is designed to help the following people: teacher, professor, student, researcher, engineer, scientist, planner, analyst, venture capitalist or inventor that is involved in one or more of the following fields:
- Power and Energy
- Physics
- Cosmology, Astronomy or Astrophysics
- Mathematics
- General science

If this describes you, then this web site is to help you stay current in your profession.

On the web site, we plan to provide the following:
- Distribute free ebooks of this report.
- Provide updates to the sections in this report.
- Correct any errors.
- Announce new discoveries.
- Help coordinate research related to TGP.
- Provide status and results of TGP experiments.
- Much more.

15.0 Conclusion

I hope you enjoyed reading this report and learned something important. If you enjoyed reading this report and went back and reread some sections, then you are probably one of the few who have a scientific curiosity. I hope you go on to make significant contributions that help mankind.

This is just the beginning. There is a lot more to be done. If you would like to see this research continued, please send a contribution to Universal Power. Go to www.universalpower.org/contributions for more information.

16.0 Answers

These are the answers to questions given at the end of previous sections. It assumes that all statements in this report are true and that the Theory of Gravity Propulsion is true.

Section 3.0 Answers

3.1   When an atom radiates a mass field, it looses some mass. (Section 3.1.1)
3.2   Yes, all atoms radiate a mass field all of the time. (Section 3.1.1)
3.3   When two mass fields meet, they create gravity propulsion fields. (Section 3.2.1)
3.4   When a mass field meets a gravity propulsion field, they create stronger gravity propulsion fields. (Section 3.2.1)
3.5   When an atom encounters a gravity propulsion field, the atom is pushed back some and its mass increases some. (Section 3.2.3)

Section 4.0 Answers

4.1   The four effects produced by a fast spinning disk are: The Deflection Effect, The Lift Effect, The Time Effect and The Time Shift Effect. (Sections 4.3.1 to 4.3.4)
4.2   Time speeds up on the outside of a spinning object. (Section 4.3.3)
4.3   Time slows down at the center axis of a spinning object. (Section 4.3.3)
4.4   John R. R. Searl invented and built the first flying saucer. (Section 4.5)
4.5   The main thing that the Podkletnov's experiment measured was the weight loss of an object held above the center of a fast spinning disk. (Section 4.6)

Section 5.0 Answers

5.1   Mass fields travel outward from a mass. (Section 5.0, Step 1)
5.2   Gravity propulsion fields travel in all directions. (Section 5.0, Step 2)
5.3   Gravity is the resultant force between a mass field and gravity propulsion fields acting on an object. (Section 5.0, Step 3)

5.4 When the earth's mass field and sun's mass field meet, they create gravity propulsion fields. (Section 5.1)

5.5 The variables that can contribute to reduce an object's weight are; D_a and I. (Section 5.6)

Section 6.0 Answers

6.1 Time is the result of all forces that make things happen. (Section 6.0)

6.2 When gravity propulsion fields are stronger, time speeds up. (Section 6.1)

6.3 Gravity slows down time. (Section 6.2)

6.4 Interference of mass fields with generated gravity propulsion fields speeds up time. (Section 6.6)

6.5 Yes. It is possible for people to increase and decrease the speed of time. For example, we can control the speed of rotating objects that deflect gravity propulsion fields. This increases the speed of time in some areas and reduces the speed of time in other areas. (Section 6.5 to 6.7)

Section 7.0 Answers

7.1 The expansion of the universe is accelerating because resultant gravity propulsion forces are constantly pushing matter outward. (Section 7.2)

7.2 Gravity propulsion forces are the strongest at the center of the universe. (Section 7.3)

7.3 In the center of the universe, the speed of time is faster. This is because gravity propulsion forces are stronger there. (Section 7.12.3)

7.4 In the center of the universe, galaxies are closer together than galaxies around us. (Section 7.12.1)

7.5 On earth, the gravity propulsion force per cubic inch is getting weaker. This is happening throughout the universe. (Section 7.4)

7.6 On earth, the mass of an average particle is getting less. As gravity propulsion fields leave a particle, the particle mass is reduced. Some gravity propulsion fields leave towards the edge of the universe and do not return to replenish the lost mass. This is happening throughout the universe. (Section 7.6)

7.7 On earth, our speed of time is slowing down. This is because gravity propulsion forces are leaving us. This is happening throughout the universe. (Section 7.4)

7.8   Our speed of light appears to be speeding up. But, actually is it our speed of time that is slowing down. (Section 7.5)

7.9   The two natural forces that increase the speed of starlight are:
1. Ether Flow. When light travels with the ether flow, away from the center of the universe, it increases the light's speed as well as its momentum, kinetic energy and effective frequency.
2. Gravity. Light traveling towards a mass speeds up because of the mass' gravity. (Section 7.10.3)

7.10   Type Ia supernovae that reach peak brightness faster then dim faster than others are because they are closer to the center of the universe. Time is faster there. (Section 7.12.3)

Section 8.0 Answers

8.1   The two wrong conclusions of the Michelson-Morley experiment were:
1. The speed of light is a constant. (It isn't a constant.)
2. Ether does not exist. (Ether does exist.) (Section 8.1)

8.2   Einstein's Theory of Special Relativity was based on the Michelson-Morley experiment. (Section 8.3)

8.3   According to The Theory of Special Relativity, the maximum speed that energy can travel is the speed of light.
According to The Theory of Gravity Propulsion, the maximum speed that energy can travel is the speed of quantum strings. The speed of quantum strings is faster than the speed of light. (Section 8.3)

8.4   According to The Theory of General Relativity, gravity is transmitted by mass warping space-time. However, it does not explain how a mass warps space-time.
According to The Theory of Gravity Propulsion, quantum strings transmit gravity. (Section 8.4)

8.5   The Theory of General Relativity does not explain how electrostatic energy and magnetic energy is transmitted.
The Theory of Gravity Propulsion explains that quantum strings transmit all types of energy including electrostatic energy and magnetic energy. (Section 8.4)

8.6   The Quantum Mechanics Theory requires some electron orbits to travel through the nucleus of an atom. The Theory of Gravity Propulsion requires that electron orbits always revolve around the nucleus. (Section 8.5)

8.7 No. The previous Unified Field Theory did not explain how gravity works. However, TGP's Quantum String Theory does explain how all forces work including gravity. (Section 8.9)

8.8 The Theory of Gravity Propulsion is a more elemental theory than The Superstring Theory. The Theory of Gravity Propulsion explains how a single type of quantum string creates all matter and energy. (Section 8.10)

8.9 The Quantum Electrodynamics Theory (QED) was initially developed to explain the Lamb Shift. (Section 8.7)

8.10 Quantum Chromodynamics Theory (QCD) states that gluons particles hold nuclei together. TGP states that surrounding gravity propulsion forces hold all particles including nuclei together. (Section 8.8)

Section 9.0 Answers

9.1 Ether is quantum strings traveling through space. (Section 9.1)

9.2 All particles radiate quantum strings. (Section 9.1)

9.3 An attractive force is when traveling strings push the particles together more than traveling strings push them apart. (Section 9.2 Attraction Forces)

9.4 Electrons are composed of more –Strings than +Strings. (Section 9.4.1)

9.5 Every string has these six properties: Polarity (of charge), Alignment, Direction, Frequency, Speed and Length. (Section 9.4.3)

9.6 When a traveling –String meets an equal traveling +String head on, and they are aligned to each, then other nothing happens. This is because they are aligned to each other. If they were not aligned, then their polarities would change. (Section 9.4.4 Principle 4C1)

9.7 When a string is deflected, its polarity changes some. (Section 9.4.4 Principle 4C3)

9.8 When a +String meets an electron, the electron absorbs it. (Section 9.4.2 Principle 2A)

9.9 Incoming strings hold particles together. (Section 9.9.1)

9.10 Anti-matter are particles with approximate equal mass and opposite string polarities. (Section 9.10)

Section 10.0 Answers

10.1 Invisible matter is composed of quantum strings with polarities that are different from ours. (Section 10.1)

10.2 We can detect invisible matter by their gravity effects on nearby visible stars. (Section 10.2)

10.3 A halo of invisible stars (HIS) is invisible stars around the rim of a visible galaxy. (Section 10.3)

10.4 The Boötes Void has invisible galaxies in it. (Section 10.4)

10.5 A halo galaxy is a halo of visible stars with a dark center of invisible stars. (Section 10.5)

10.6 The Lamb Shift is caused by invisible particles that help push electron orbits outward. (Section 10.6)

10.7 Invisible binary stars are a theoretical rare possibility where two stars, which are invisible to each other, are closely revolving around each other. (Section 10.7)

10.8 The Continuity Principle says that the only thing that can exist is the final steady state timeline. (Section 10.8)

10.9 The zero dimension created the Big Bang. (Section 10.8)

10.10 Quantum strings create all matter, energy, fields, forces, space and time. (Section 10.8)

Section 11.0 Answers

11.1 A gravity transmitter transmits gravity signals by vibrating a mass. (Section 11.0)

11.2 The two types of gravity transmitters are: Magnetic and Electrostatic. (Section 11.1)

11.3 The problem with gravity transmitters is that they can only broadcast in all directions. Radio transmitters can focus their broadcast power in specific directions. (Section 11.1)

11.4 The two types of gravity receivers are: Magnetic and Optical. (Section 11.2)

11.5 A signal from a gravity transmitter travels faster than a signal from a laser. (Section 11.3)

Answer To The Final Question

Here are the facts presented in this book that support the Theory of Gravity Propulsion.

Theory of Gravity Propulsion Predictions	Facts
TGP explains how one type of quantum string creates all of the forces listed to the right.	The following forces exist: Gravity forces. Electromagnetic, light and photon forces. Strong force that holds all particles together. Weak forces. Newton's Laws of Motion. Centrifugal force. Electrostatic forces among electrons, protons and neutrons. Particle magnetic poles forces. Lamb shift forces.
TGP explains how the same type of quantum string creates the different types of matter listed to the right.	Matter, subatomic particles, anti-matter and invisible (dark) matter exist.
Mass and energy are composed of the same type of quantum string.	Mass and energy are related.
There is a center of the universe. In the center of the universe, time goes fastest. The closer things are to the center, the faster time goes.	Type Ia supernovae show that time goes faster the further away they are from us towards the direction of Virgo.
Gravity propulsion fields are pushing all matter outwards.	Our universe is expanding.
Gravity propulsion fields are constantly pushing and accelerating all matter outward.	The expansion of our universe is accelerating.
There is a center of the universe. In the center of the universe, galaxies are the densest.	There is an area in our universe where galaxy clusters are the densest. This is called the Virgo cluster of galaxies.

In the direction of the center of the universe, galaxies are denser and time goes faster. Therefore they will be a supercluster of galaxies that appears closest to us. And, it will be in the direction towards the center of the universe.	There is a local supercluster of galaxies. It is in the direction of Virgo.
There is reduced weight above the center of a spinning object because gravity propulsion fields are deflected from that area.	Podkletnov's experiment showed reduced weight above a spinning object.
If you spin a cylinder fast enough it will overcome the force of gravity and lift up.	Searl's Levity Discs use fast rotating cylinders to make them lift up.
The further you are from gravity, the stronger gravity propulsion fields are and the faster time goes.	Gravity Probe A experiment shows that time speeds up the further you are from gravity.
There are invisible stars surrounding visible galaxies. Their gravity effects cause outermost visible stars to rotate around the galaxy at closer to the rate that the innermost stars do.	The outermost stars of a galaxy revolve around the galaxy at about the same speed as the innermost stars.

There is no other theory that can explain all of the above facts. The only thing that TGP contradicts with is other theories. TGP does not contradict any facts.

This report has identified experiments that can verify or disprove 18 additional and unique TGP predictions. These experiments will establish new facts.

Appendix A: Creating the Cosmological Equation

Phase 1

This section defines the statement of the cosmological equation to solve. This should be a good project for a mathematics graduate student. Here is the statement of the problem.

Initial State
- Mass: Start with a solid sphere (ball) of 100% mass "M". M = 1. It has a radius R = 1. This is the only measure for distance. The mass is a collection of unknown number of atoms. This represents the initial mass of the universe. This initial density absorbs all gravity propulsion forces. As this density is reduced, it proportionately lets gravity propulsion forces through.
- Force: Start with a fixed amount of force F. F = d * M. d is a constant. F is equally distributed among the mass. F is radiated out around each atom in all directions. This represents initial gravity propulsion forces.
- Speed: F travels the distance R in one T unit of time. R is the radius. T is one unit of time. Speed of F = R/T. This represents the fixed speed of gravity propulsion fields. This assumes that the speed of gravity propulsion fields is a constant.
- Interaction: When F hits a mass (a subset of the initial mass), it moves the mass in that direction if it can. (F = MA).

Provide equations for the following:

Equation #1: Change of F (dF) as a function of change of R (dR) given R and T. This means how much energy (F) leaves a mass minus how much energy (F) the mass absorbs depending on where (R) and when (T) it is. Mass is function of dR given R and T. M = f(dR, R, T) dF = f(dR, R,T).

This equation shows how fast energy leaves mass over time throughout the model universe. This is needed for phase 2 to determine how much energy is converted from mass as atoms slow down.

Equation #2: The speed of mass (R/T), at any point, outward from the center as a function of R and T. Speed = f(R, T)
This equation shows relative speeds of mass and light in the model universe. This is needed for phase 2 to determine the speed of light relative to the universe.

Equation #3: Total F force (from all directions) on a mass, at any point, as a function of R and T. F = f(R, T)
Since F = relative time, this equation shows the relative speed of time at any point in the model universe. This is needed for phase 2 to determine the relative speed of time and light (C) relative to any point.

Creating the Cosmological Equation: Phase 2

The phase 1 cosmological equation has limitations. It is a one shot deal. It shows the effects that an initial amount of F has on matter over time and space (speed and distance). It does not show what happens as matter is converted into more energy (F) as things slow down. Phase 2 is to add in the "mass reduction effect" to this fixed model and determine the new three equations.

Equation #1 of phase 1 shows how fast energy and mass is lost over time throughout the model universe. This lost energy comes from the mass of the atom.

Equation #2 of phase 1 shows how the speed of light changes relative to the model universe's time (T) and distance (R).

Equation #3 shows the relative speed of time at any point in the model universe. Together with equation #2 we can know the speed of time and light (C) relative to any point. C is a function of R and T. We can then apply $F = MC^2$ to Equation #1 to determine how much mass is lost over time and space. As mass decreases, their ability to absorb gravity propulsion fields (mass density) proportionally decreases too. This affects all of the equations.

Appendix B: Derivation of Tensile Strength Equation

This appendix derives the equation for tensile strength verses revolutions per minute verses radius for a spinning cylinder. For a given material's tensile strength, you can determine the maximum revolutions per second verses radius of the cylinder. Tensile strength has to be stronger than the total centrifugal force pushing on the outside of the cylinder. Otherwise, the centrifugal force will pull the cylinder apart.

Tensile Strength > Total Centrifugal Force / Cylinder Outside Area

The variables used are defined below.
Tensile Strength of material (lb/in^2 or kg/cm^2)
D = Density of material (lb/in^3 or kg/cm^3)
Π = Pie = 3.1415926
R = Revolutions per second
r = Radius of cylinder (in or cm)
m = Mass
a = Acceleration
h = Height
dr = Delta radius

Total Centrifugal Force = Mass x Acceleration

Mass = Density (D) x Height (h) x sum of ($\int$) circumference ($2\Pi r$) x change of radius (dr)
$m = Dh2\Pi\int r dr$

Acceleration = $velocity^2$/radius
Velocity is the circumference ($2\Pi r$) times the revolutions per second (R) = $2\Pi rR$
$a = (2\Pi rR)^2/r$
$a = 4\Pi^2 r^2 R^2/r$
$a = 4\Pi^2 rR^2$

Total Centrifugal Force = m x a
                     Substituting for m and a.
                     $= Dh2\Pi\int r dr \times 4\Pi^2 rR^2$
                     $= 8Dh\Pi^3 R^2\int r^2 dr$
$\int r^2 dr = r^3/3$ therefore
Total Centrifugal Force = $(8/3)Dh\Pi^3 R^2 r^3$

Cylinder Outside Area = Circumference (2Πr) x Height (h) = 2Πrh
Tensile Strength > Total Centrifugal Force / Cylinder Outside Area
Tensile Strength > $(8/3)Dh\Pi^3R^2r^3/2\Pi rh$
Tensile Strength > $(4/3)D\Pi^2R^2r^2$

Index Table